Cells and Biomaterials for Intervertebral Disc Regeneration

Synthesis Lectures on Tissue Engineering

Editor

Kyriacos A. Athanasiou, *University of California, Davis*

The Synthesis Lectures on Tissue Engineering series will publish concise books on aspects of a field that holds so much promise for providing solutions to some of the most difficult problems of tissue repair, healing, and regeneration. The field of Tissue Engineering straddles biology, medicine, and engineering, and it is this multi-disciplinary nature that is bound to revolutionize treatment of a plethora of tissue and organ problems. Central to Tissue Engineering is the use of living cells with a variety of biochemical or biophysical stimuli to alter or maximize cellular functions and responses. However, in addition to its therapeutic potentials, this field is making significant strides in providing diagnostic tools. Each book in the Series will be a self-contained treatise on one subject, authored by leading experts. Books will be approximately 65-125 pages. Topics will include 1) Tissue Engineering knowledge on particular tissues or organs (e.g., articular cartilage, liver), but also on 2) methodologies and protocols, as well as 3) the main actors in Tissue Engineering paradigms, such as cells, biomolecules, biomaterials, biomechanics, and engineering design. This Series is intended to be the first comprehensive series of books in this exciting area.

Cells and Biomaterials for Intervertebral Disc Regeneration
Sibylle Grad, Mauro Alini, David Eglin, Daisuke Sakai, Joji Mochida, Sunil Mahor, Estelle Collin, Biraja Dash, and Abhay Pandit
2010

Fundamental Biomechanics in Bone Tissue Engineering
X. Wang, J.S. Nyman, X. Dong, H. Leng, and M. Reyes
2010

Articular Cartilage Tissue Engineering
Kyriacos A. Athanasiou, Eric M. Darling, and Jerry C. Hu
2009

Tissue Engineering of Temporomandibular Joint Cartilage
Kyriacos A. Athanasiou, Alejandro J. Almarza, Michael S. Detamore, and Kerem N. Kalpakci
2009

Engineering the Knee Meniscus
Kyriacos A. Athanasiou and Johannah Sanchez-Adams
2009

Cells and Biomaterials for Intervertebral Disc Regeneration
Sibylle Grad, Mauro Alini, David Eglin, Daisuke Sakai, Joji Mochida, Sunil Mahor, Estelle Collin, Biraja Dash, and Abhay Pandit

ISBN: 978-3-031-01452-9   paperback
ISBN: 978-3-031-02580-8   ebook

DOI 10.1007/978-3-031-02580-8

A Publication in the Springer Nature series
SYNTHESIS LECTURES ON ADVANCES IN AUTOMOTIVE TECHNOLOGY

Lecture #5
Series Editor: Kyriacos A. Athanasiou, *University of California, Davis*
Series ISSN
Synthesis Lectures on Tissue Engineering
Print 1944-0316   Electronic 1944-0308

Cells and Biomaterials for Intervertebral Disc Regeneration

Sibylle Grad, Mauro Alini, and David Eglin
AO Research Institute Davos, Switzerland

Daisuke Sakai and Joji Mochida
Tokai University, Japan

Sunil Mahor, Estelle Collin, Biraja Dash, and Abhay Pandit
National University of Ireland, Galway, Ireland

SYNTHESIS LECTURES ON TISSUE ENGINEERING #5

ABSTRACT

Disorders related to the intervertebral disc (IVD) are common causes of morbidity and of severe life quality deterioration. IVD degeneration, although in many cases asymptomatic, is often the origin of painful neck and back diseases. In Western societies IVD related pain and disability account for enormous health care costs as a result of work absenteeism and thus lost production, disability benefits, medical and insurance expenses. Although only a small percentage of patients with disc disorders finally will undergo surgery, spinal surgery has been one of the fastest growing disciplines in the musculoskeletal field in recent years. Nevertheless, current treatment options are still a matter of controversial discussion. In particular, they hardly can restore normal spine biomechanics and prevent degeneration of adjacent tissues.

While degeneration affects all areas of the IVD, the most constant and noticeable changes occur in the gel-like central part, the nucleus pulposus (NP). Recent emphasis has therefore been put in biological ways to regenerate the NP; however, there are a number of obstacles to overcome, considering the exceptional biological and biomechanical environment of this tissue. Different biological approaches such as molecular, gene, and cell based therapies have been investigated and have shown promising results in both *in vitro* and *in vivo* studies. Nonetheless, considerable hurdles still exist in their application for IVD regeneration in human patients. The choice of the cells and the choice of the cell carrier suitable for implantation pose major challenges for research and development activities. This lecture recapitulates the basics of IVD structure, function, and degeneration mechanisms. The first part reviews the recent progress in the field of disc and stem cell based regenerative approaches. In the second part, most appropriate biomaterials that have been evaluated as cell or molecule carrier to cope with degenerative disc disease are outlined. The potential and limitations of cell- and biomaterial-based treatment strategies and perspectives for future clinical applications are discussed.

KEYWORDS

intervertebral disc, degeneration, nucleus pulposus, cell therapy, mesenchymal stem cells, biopolymers, injectable hydrogel

Contents

CHAPTER 1

Cell Therapy for Nucleus Pulposus Regeneration

Sibylle Grad and Mauro Alini
AO Research Institute Davos, Switzerland

Daisuke Sakai and Joji Mochida
Tokai University, Japan

1.1 INTRODUCTION

Disorders related to the intervertebral disc (IVD) are common causes of morbidity and severe life quality deterioration. Disc degeneration, although in many cases asymptomatic, is often the origin of painful neck and back diseases. It is also associated with sciatica and disc herniation or prolapse. Due to alterations in disc height and mechanics of the spine, the surrounding muscles and ligaments may also be affected. In the long term, degenerative changes may lead to spinal stenosis, which is a significant health problem in elderly people. With the current demographic changes and an increasing proportion of the aged population, the incidence of disc related disabilities is rapidly rising (Frymoyer, J., 1992; Rubin, D., 2007).

Compared to other musculoskeletal tissues, the IVD degenerates early in life, with first signs of degeneration identified already in the adolescent age group (Boos et al., 2002). With aging, the degree of degeneration increases steadily so that around 10% of the 50-year old and 60% of the 70-year old suffer from severely degenerate discs (Miller et al., 1988). In fact, nearly 12 million Americans are diagnosed with IVD diseases each year (Wang et al., 2007a). In Western societies, IVD related pain and disability account for enormous health care costs as a result of work absenteeism and thus lost production, disability benefits, medical and insurance expenses. The lifetime prevalence of spinal pain has been reported as 54% to 80%, and studies of the prevalence of low back pain and neck pain have revealed 23% of patients reporting grade II to IV low back pain (high pain intensity with disability) versus 15% with neck pain (Manchikanti et al., 2009).

Treatment usually begins with non-operative modalities, such as physical therapy or methods for core strengthening; symptomatic medical treatment with non-steroidal anti-inflammatory medications is a further common method to reduce pain. Surgical intervention is considered if conservative therapy fails. Although only a small percentage of patients with disc disorders finally will undergo surgery, spinal surgery has been one of the fastest growing disciplines in the musculoskeletal field in recent years (Deyo and Mirza, 2006; Gray et al., 2006). Nevertheless, current treatment options are still a matter of controversial discussion. The standard surgical intervention has been spinal arthrodesis with the aim to immobilize the spinal segment, preferably by bony fusion (Bono and Lee, 2004; Gibson and Waddell, 2005). The aim is to cease mechanical cues and inflammatory processes causing pain and disability. However, compared to conservative treatment only small benefits could be achieved, as assessed in several clinical studies (Brox et al., 2006; Fairbank et al., 2005). In addition, the occurrence of adjacent segment disease should not be underestimated. With the goal to better preserve the biomechanics of the spine, total disc replacement has been introduced and has become part of surgical routine in recent years. Since this technology has not been able to demonstrate any significant advantage to the standard spinal arthrodesis and in contrast has faced considerable complication rates, it has been critically debated (Blumenthal et al., 2005; Khan and Stirling, 2007). Other newer technologies include nucleus pulposus replacement or dynamic stabilization methods (Di Martino et al., 2005; Nockels, R., 2005). Long-term clinical outcomes that may disclose any potential benefit of these new methods are not yet available; however, the success rates of all these procedures are generally similar.

Although satisfying results can be achieved, all of these treatment methods only attempt to reduce the pain, but they cannot repair the degenerated disc. In particular, they hardly can restore normal spine biomechanics and prevent degeneration of adjacent tissues. Therefore, new treatments are under development with the aim to restore disc height and biomechanical function. The objective of such regenerative strategies is to generate healthy disc tissue or functional replacements that decelerate or reverse painful degeneration processes. A number of biological approaches such as molecular, gene, and cell based therapies have been investigated and have shown promising results in both in vitro and in vivo studies. Nonetheless, considerable hurdles still exist in their application for IVD regeneration in human patients.

While degeneration affects all areas of the IVD, the most constant and noticeable changes occur in the nucleus pulposus (NP), the gel-like central part. The loss of NP cells has been correlated with the onset of disc degeneration, leading to the inability to maintain the functional load-absorbing extracellular matrix with concomitant fibrous transformation. As a result, much emphasis has been put in biological ways to regenerate the NP. Indeed, with its relatively simple matrix structure, engineering new NP tissue may be considered a relatively easy and straightforward task. However, there are a number of obstacles to overcome, given that the tissue is maintained in an exceptional biological, biophysical, and biomechanical environment. The choice of the cell population and the cell carrier suitable for implantation pose major challenges for research and development activities. In this first part of this lecture, we will review the recent progress in the field of cell based regenerative approaches, while the second part will introduce most appropriate biomaterials that have been evaluated as cell or molecule carriers to cope with degenerating disc disease.

1.2   INTERVERTEBRAL DISC DEGENERATION

1.2.1   NORMAL DISC STRUCTURE AND FUNCTION

The IVDs are located between the vertebral bodies and are composed of three morphologically distinct regions: The NP is the central part and contains randomly organized collagen fibers and elastin fibers that are embedded in a highly hydrated gel-like matrix rich in proteoglycan, with aggrecan as the major component. The cells of the NP are often referred to as chondrocyte-like because of their morphology and phenotype that is similar to the one of articular chondrocytes. These cells are interdispersed at low density (approximately 5000 cells/mm³ (Maroudas et al., 1975)) and sometimes arranged in clusters within the matrix. Surrounding the NP is the annulus fibrosus (AF) which is composed of 15-25 concentric rings or lamellae (Marchand and Ahmed, 1990). Collagen fibers are arranged parallel within each lamella, but they alternate between adjacent lamellae and are oriented at approximately 60° to the vertical axis. Elastin and proteoglycans are present between the lamellae that support the maintenance of the disc shape after bending, i.e., flexion or extension (Yu, J., 2002). The cells of the AF are morphologically and phenotypically similar to fibroblasts; they are thin and elongated and aligned parallel to the collagen fibers. The cell density is slightly higher than in the NP (approximately 9000 cells/mm³). The third morphologically distinct site of the IVD is the cartilaginous endplate, a thin layer of hyaline cartilage located between the disc and the vertebral body. Collagen fibers run parallel to the vertebral bodies.

The function of the IVD is to provide flexibility and shock absorbing capabilities to the spinal column. The compressibility of the NP aids to evenly distribute the loads arising upon the vertebral bodies during motion activities that include flexion and extension. The mechanical functions are provided by the specialized extracellular matrix. The network of collagen type I, which makes about 70% of the dry weight of the AF, and collagen type II, which is found in the NP and inner AF, together with other collagens, mainly types III, V, VI, IX, XI, provides tensile strength to the disc and serves as an anchor to the vertebral bone. The NP contains large proportions of hyaluronan and proteoglycan, mostly aggrecan, although other proteoglycans such as biglycan, decorin, and fibromodulin are also present. The high osmotic pressure provided by the sulfated glycosaminoglycan side chains (chondroitin sulfate and keratan sulfate) is important to maintain tissue hydration and absorb compression forces.

The matrix of the IVD is a dynamic structure, whereby the cells ought to maintain disc homeostasis by slow but constant breaking down of matrix molecules and forming new matrix components. Half-lives of aggrecan were found to be about 12 years in normal and about 8 years in degenerated human discs, while the half-life of collagen even exceeds 90 years (Sivan et al., 2006, 2008). Synthesis of factors involved in the complex anabolic and catabolic processes are thoroughly balanced within functional tissues. Important anabolic factors that promote matrix formation are growth factors, primarily transforming growth factor-beta and members of its superfamily such as the bone morphogenetic proteins, growth and differentiation factor 5, and insulin-like growth factor. Anti-catabolic factors like tissue inhibitors of metalloproteinases (TIMPs), general proteinase inhibitors, and anti-inflammatory molecules also contribute to the maintenance of tissue homeostasis. On the other hand, matrix degradation is mediated by catabolic enzymes and inflammatory cytokines like interleukin-1 and tumor necrosis factor-alpha (Kang et al., 1997; Le Maitre et al., 2004; Seguin et al., 2005). Proteinases, particularly members of the matrix metalloproteinase (MMP) family are involved in the breakdown of matrix molecules. These enzymes have been documented to degrade collagens (MMP1, MMP8, MMP13), gelatins (MMP2, MMP9), and other macromolecules (MMP3, MMP7). Furthermore, aggrecanases of the ADAMTS (a disintegrin and metalloprotease with thrombospondin motifs) family may play a major role in the proteoglycan turnover of the IVD (Le Maitre et al., 2004).

The metabolism of the IVD is determined by its avascular nature. Nutrients and metabolites can reach the disc essentially by diffusion through the vertebral endplates and the AF. As a result, oxygen tension within the disc is significantly reduced and the disc cell metabolism is partly anaerobic, which leads to high concentrations of lactic acid and low pH conditions (Grunhagen et al., 2006; Urban et al., 2004).

1.2.2   AGING AND DEGENERATION

There are still no clear and widely recognized definitions for disc degeneration. It has been generally suggested that disc degeneration may mimic the age-related changes but occurs at an accelerated rate (Adams and Roughley, 2006; Le Maitre et al., 2007); hence disc degeneration seems to be a process of premature aging.

Basically, disc degeneration can result from an imbalance between anabolic and catabolic processes or simply from a loss of the steady-state metabolism that is maintained in the normal healthy disc. A deceleration of matrix turnover has been shown to cause increased cross-linking of collagen fibers and disc stiffening (Duance et al., 1998; Pokharna and Phillips, 1998). The balance between catabolism and anabolism can be altered by a number of factors including metabolic changes as a result of altered nutrition, biomechanical cues, genetic predisposition, or a combination thereof. When a shift towards catabolism occurs, a loss of proteoglycan in the NP is the most relevant biochemical issue. Large aggrecan molecules are degraded and small fragments may be released from the tissue (Sztrolovics et al., 1997). A decrease in the osmotic pressure occurs with a concomitant loss of hydration, impairing the mechanical functionality of the NP. The population of the small proteoglycans also changes with disc degeneration. Relative increases in the amount of decorin and particularly biglycan were reported in degenerate human discs (Inkinen et al., 1998). It was also found that specific changes in the concentration of distinct matrix molecules occurred in the human disc in an age-related manner, whereby the rise in the level of biglycan was corroborated in both the NP and AF compartments (Singh et al., 2009). Moreover, fibronectin increases with degeneration and it becomes more fragmented (Oegema et al., 2000); fibronectin fragments, in turn, are known to activate MMP synthesis and inhibit aggrecan production, accelerating the degradation process. Synthesis and composition of collagens also vary, with enhanced type II collagen synthesis in the NP during early stages of degeneration and enhanced type II collagen appearing in the AF in more advanced degeneration. At the same time, type I collagen is increasingly produced in the NP, leading to fibrotic transformation of the NP tissue and progressive inability to identify a clear border between the NP and AF tissues. In addition, while the fibrillar collagens, particularly type II collagen, become more denatured, type X collagen has been localized in degenerate discs in association with clusters of chondrocytic cells (Antoniou et al., 1996; Boos et al., 1997).

Matrix degradation processes are mostly mediated by MMPs and ADAMTS. The expression and activity of a number of MMPs, including MMP 1, 3, 7, 9, and 13 are increased during disc degeneration (Le Maitre et al., 2004, 2006; Roberts et al., 2000; Weiler et al., 2002). Since TIMP 1 and 2 have also been found up-regulated in degenerate discs, MMPs that are not suppressed by these specific inhibitors appear to play a more significant role in the pathogenesis of the IVD. In addition, a number of enzymes with aggrecanase activity, such as ADAMTS 1, 4, 5, 9, and 15 have been reported with increased gene and protein levels during disc degeneration, without up-regulation of their inhibitor TIMP 3 (Le Maitre et al., 2004; Patel et al., 2007; Pockert et al., 2009).

As a result of decreased proteoglycan in the NP, the hydrostatic pressure lowers and the disc space begins to collapse. The supporting AF and the ligaments lose tension and start to buckle and fissure. Spinal micro-instability that can develop is associated with abnormal loading of the facets with circumferential degenerative and, eventually, osteoarthritic or hypertrophic changes (Butler et al., 1990). The cascade of degenerative processes may lead to further instability, inflammatory reactions, and potentially disc herniation. Concomitant with matrix degradation and reduced disc height is often an in-growth of blood vessels and nerves into the normally avascular and aneural tissue (Freemont et al., 1997). The latter can be an effect of proteoglycan decrease since aggrecan has been shown to inhibit neural in-growth into the disc (Johnson et al., 2002). All these changes are strongly associated with back pain, radiculopathy, and myelopathy.

As mentioned above, the predominant factors that may disturb IVD homeostasis are of metabolic, mechanical, and genetic origin. A decrease in the nutrient supply to the disc cells is considered to be a major cause of failure. Nucleus cells are supplied by capillaries that originate in the vertebral body, penetrate the subchondral plate, and terminate above the cartilaginous endplate. Nutrients thus need to diffuse through the endplate and the extracellular matrix to the NP cells, which may be as far as 8 mm from the capillary bed (Holm et al., 1981). Consequently, smoking, atherosclerosis, and other diseases that affect the microvasculature of the vertebral body can accelerate degenerative changes of the IVD. Finally, even if the blood supply to the vertebral body is sufficient, nutrients may not reach the disc cells if the cartilaginous endplate permeability is impaired, e.g., as a result of calcification processes (Roberts et al., 1996). Associations between endplate cartilage damage and diffusion into the disc and a high correlation between the density of openings in the osseous endplate, particularly of the size of the capillary buds, and the morphologic degeneration grade of the disc have been described (Benneker et al., 2005; Rajasekaran et al., 2004).

Non-physiological mechanical load has also been shown to provoke changes leading to disc degeneration. It was suggested for many decades that excessive loading, often work-related, is a major cause of disc-related low back pain. In fact, high frequency load resembling whole body vibration has been associated with disc disorders, and these findings have recently been complemented by a clinical study reporting that whole-body vibration was a significant determinant of severe disc degeneration (Kuisma et al., 2008; Pope et al., 1998). Evidence for a role of adverse mechanical forces is also provided by the frequent occurrence of adjacent disc disease following spinal fusion (Eck et al., 1999). Experimental overloading or mechanical injury has been shown to induce degenerative changes in animal models (Iatridis et al., 2006; Osti et al., 1990) and to affect disc cell viability in whole IVD organ culture (Illien-Junger et al., 2010). However, intense exercise does not appear to have adverse effects on the discs (Puustjarvi et al., 1993). Moreover, several in vivo and in vitro studies failed to observe significant effects of mechanical cues on the IVD structure, and it has become increasingly obvious that environmental factors such as occupation show little correlation with the incidence of disc degeneration (Battie and Videman, 2006).

The significance of the genetic predisposition has been discovered more recently. Several studies reported a strong familiar association with the frequency of disc diseases, and in twin studies a heritability of more than 60% was documented (Battie et al., 1995; Matsui et al., 1998; Sambrook et al., 1999; Varlotta et al., 1991). Gene polymorphisms of a number of extracellular matrix macromolecules have been associated with an increased susceptibility for IVD degeneration. Mutations in collagen types I, IX, XI, aggrecan core protein, cartilage intermediate layer protein (CILP), and asporin genes have been found to correlate with disc degeneration in different populations. Studies on transgenic mice have confirmed that mutations in molecules that affect the properties of the extracellular matrix may lead to degenerative changes in the IVD. Furthermore, polymorphisms in the vitamin D receptor and in the MMP 3 gene promoter region were associated with disc degeneration (Kalichman and Hunter, 2008). This shows evidence that mutations in various gene classes can initiate alterations in the biochemistry, structure, and function of the disc. Identification of the genes involved and of the respective polymorphisms may make new diagnostic criteria possible. However, on account of the apparent multifactorial nature of disc degeneration and the evidence for interaction between genetic and environmental components isolated genetic investigations are doubtful to serve as a diagnostic tool. Complementing new imaging methods that can detect minor matrix changes (Mwale et al., 2008), specific genotyping may though aid in recognizing patients with early degenerative signs. These patients in their early stages of pathological disc degeneration may benefit from regenerative biological treatment methods described in the following paragraphs (Fig. 1.1).

1.3   CELL BASED NUCLEUS PULPOSUS REGENERATION

1.3.1   DISC CELL BASED APPROACHES

The process of disc degeneration has been defined as "aberrant cell-mediated response to progressive structural failure" (Adams and Roughley, 2006), and, in fact, the disc cells undergo significant changes during the course of this process. Most remarkably, the cell type of the NP substantially changes in early life. During development of the IVD, the NP is primarily populated by notochordal cells, which may persist throughout most of adult life in certain species such as mouse, rat, rabbit, porcine, and some canine breeds. In other species including human beings, the number of noto-

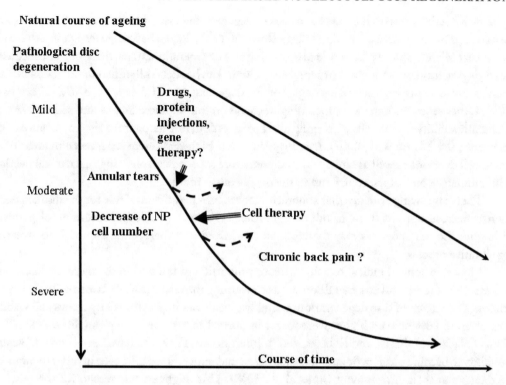

Figure 1.1: Schematic of the time course of normal aging versus pathological disc degeneration. Biological therapies, including cell therapy, gene therapy, drugs, or protein injection may have the potential to reverse the progression of degeneration by the appropriate timing of intervention. Chronic back pain may be associated to the period where pathological disc degeneration progresses from moderate to severe. (Source: Sakai, D., Future perspectives of cell-based therapy for intervertebral disc disease, *European Spine Journal* 17 Supp.4:452–458, 2008.)

chordal cells decreases rapidly after birth, and they eventually disappear within the first decade of life (Pazzaglia et al., 1989). Notochordal cells are more effective in the synthesis of proteoglycans than the chondrocyte-like cells found in adult NP and are thus essential for the maintenance of the gel-like matrix of healthy NP tissue (Cappello et al., 2006). Moreover, notochordal cells have been shown to stimulate proteoglycan synthesis of other NP cells (Aguiar et al., 1999; Erwin and Inman, 2006). While the clinical feasibility of notochordal cell implantation is doubtful, specific administration of soluble factors produced by these cells or even strategies to maintain this cell population beyond childhood hold great promise.

It has widely been recognized that there is a continuing increase in cell death during disc aging and degeneration (Boos et al., 2002; Gruber and Hanley, 1998). On the other hand, enhanced proliferative activity has been observed in cells of degenerating discs, possibly as a repair reaction to the destruction and failure of the disc matrix (Johnson et al., 2001). Several investigations have also shown that cell senescence plays a role in disc aging and degeneration. In particular, senescence associated β-galactosidase, an important senescence biomarker, has been identified in herniated NP and significantly increases with increasing degree of disc degeneration (Gruber et al., 2007; Roberts et al., 2006). Other senescence markers, including decreased mean telomere length and reduced *in vitro* proliferative ability of disc cells with aging and degeneration, further confirm the importance of cell senescence (Le Maitre et al., 2007). It is suggested that both replicative senescence in areas of repeated cell divisions as well as stress-induced premature senescence, following mechanical overload or inflammatory mediators, contribute to the degenerative process.

There is growing evidence that apoptosis mechanisms play a central role for the disc cell death, whereas necrosis appears to be of minor significance (Zhao et al., 2006). Prevention of pathways leading to disc cell apoptosis may therefore be considered as a therapeutic approach to arrest the degenerative process.

Finally, in spite of indications that a major proportion of the cells in the mature adult human IVD are dead, there is still no general consensus concerning the relationship between cell viability and either age or degree of disc degeneration. Using live/dead viability/cytotoxicity assays cell viability rates of up to 90% or even higher were found in normal human disc samples (Bibby et al., 2002; Chen et al., 2005; Johnson and Roberts, 2007). Interestingly, IVD cells have been shown to acquire the ability to become phagocytic and may thus ingest and remove apoptotic cells in order to minimize the damage to their environment (Jones et al., 2008). This might be one reason for the relatively high cell viability rate observed despite the well-documented occurrence of apoptotic cell death in the IVD.

1.3.1.1 In Vitro NP Cell Culture and Tissue Engineering Studies

Given that the degenerative cascade is characterized by a loss of viable cells and a change in the metabolic characteristics of the living cells, any biologic treatment must aim to enhance the number of viable cells and assure the production of the appropriate matrix components. Besides growth factor injection and gene transfer, cell implantation is of particular therapeutic potential since it addresses both the depleted cell population and the maintenance or restoration of the proper matrix. The ideal candidate for cell implantation may be the disc cell itself since disc cells are expected to easily adapt to the environment and to synthesize the correct matrix molecules. Many in vitro studies have been performed to optimize conditions for tissue engineering of IVD and specifically of NP tissue since disc degeneration is believed to originate in the NP region. Two environmental factors that need to be considered for NP cells are nutrition and mechanical load. Mature NP cells appear to generate ATP primarily through glycolysis, thereby using glucose and producing lactic acid at a high rate (Bibby et al., 2005). Cell metabolic activity is reduced under low oxygen tension, although the cells do not seem to require oxygen to remain viable (Horner and Urban, 2001). However, in vitro cell and organ culture studies have revealed that the NP cell survival is impaired when concentrations of glucose fall below physiological levels (Bibby and Urban, 2004; Junger et al., 2009). Low pH values (pH<6.7) that may arise from accumulation of lactic acid have also been shown to affect NP cell vitality (Bibby and Urban, 2004).

Like articular chondrocytes NP cells undergo rapid phenotypical changes when they are cultured in monolayer conditions, as evidenced by decreases in gene expression of type II collagen, aggrecan, and reduced proteoglycan synthesis rate (Horner et al., 2002; Kluba et al., 2005). At least partial redifferentiation can be achieved when cells are transferred to a suitable three-dimensional system. Three-dimensional scaffolds not only help to maintain or re-gain the desirable cell phenotype but also provide appropriate mechanical environment and potentially biochemical signals. A variety of biomaterials have been used as cell carrier systems for NP tissue engineering, including alginate, chitosan and its derivatives, collagen and atelocollagen, gelatin, hyaluronan formulations, polylactic acid, poly(lactic-co-glycolic) acid, calcium polyphosphate, fibrin, and polyurethane (O'Halloran and Pandit, 2007). The prospective of the different materials is discussed in the second part of this lecture.

Under physiological conditions, the IVD is always under load that arises from body weight and from muscle activity. The pressure in the human lumbar disc has been reported to be around 0.2 MPa when lying down and may increase 3 to 5-fold upon rising depending on posture and activity (Wilke et al., 1999). Numerous studies have investigated the effect of mechanical stress on NP cells in vitro (Setton and Chen, 2004). For example, cyclic tensile and shear stresses at 0.05 Hz frequency resulted in an up-regulation of DNA and collagen synthesis in rabbit NP cells (Matsumoto et al., 1999). The response of IVD cells to unconfined uniaxial compression was investigated in a rabbit explant model (Wang et al., 2007b). Static compression at 0.5 MPa and 1 MPa and dynamic compression at 0.5 MPa and 1 MPa and frequencies of 0.1 Hz and 1 Hz were applied. Static compressive load suppressed anabolic gene expression, whereas under dynamic compression the disc underwent significant anabolic changes with up-regulation of aggrecan and collagen types I and II gene expression. Furthermore, increases in interleukin-1-beta and tumor necrosis factor-alpha expression and TUNEL positive cells were noted under all loading conditions, with most pronounced effects in statically loaded explants. This corroborates that static compression has a catabolic role, while dynamic load at appropriate level may promote the synthetic activity of the disc. Hydrostatic pressure has widely been applied, given that this type of load naturally occurs in the disc in vivo. In canine lumbar NP cells embedded in alginate, up-regulation of aggrecan and collagen gene expression and increases in collagen and proteoglycan synthesis could be observed upon application of hydrostatic pressure at 1 MPa (Hutton et al., 1999). Furthermore, short-term high amplitude and frequency hydrostatic loading stimulated protein synthesis and inhibited degradation in an alginate culture system of rabbit NP cells (Kasra et al., 2003). On the other hand, a reduced synthesis rate and increased degradation was observed in porcine NP cells near the 5 Hz frequency upon application of hydrostatic pressure at 1 MPa and different frequencies (Kasra et al., 2006). The influence of hydrostatic pressure appears to be complex and both frequency and magnitude dependent. For NP cells seeded in collagen type I matrices, it was reported that low hydrostatic pressure

(0.25 MPa, 30 min, 0.1 Hz) has minimal effects with a tendency to anabolic reaction, whereas high hydrostatic pressure (2.5 MPa, 30 min, 0.1 Hz) appears to decrease the matrix protein expression with a tendency to increase some matrix-turnover enzymes (Neidlinger-Wilke et al., 2006). Furthermore, the effect of mechanical load also needs to be viewed in the context of other surrounding conditions, such as pH, oxygen tension, and osmolarity. For example, extracellular osmolarity has been shown to influence the response of NP cells to mechanical stimuli (Wuertz et al., 2007). These observations point to the fact that mechanical stimuli may be used for in vitro conditioning of isolated IVD cells; however, care should be taken in the translation of the outcome to the in vivo situation due to the different environment. It was documented that in the NP and inner AF of bovine discs, application of hydrostatic pressure in the range of 1-7.5 MPa for only 20 seconds stimulated matrix synthesis, with maximal stimulation after application of 2.5 MPa. Exposure to 2.5 MPa also stimulated synthesis in human NP, while exposure to 10 MPa inhibited proteoglycan synthesis (Ishihara et al., 1996).

Disc cell metabolism is modulated by a variety of soluble factors that act in a paracrine or autocrine manner. These factors can increase the synthesis of extracellular matrix components, block their catabolic breakdown, or affect both the matrix production and degradation. Previous studies have documented beneficial effects of a number of growth factors on NP cells in culture (O'Halloran and Pandit, 2007). Growth factors can be applied via delivery of recombinant pure or "embedded" proteins or as a prolonged supplement by gene therapy. Transforming growth factor-beta (TGF-beta), insulin-like growth factor 1 (IGF-1) platelet-derived growth factor (PDGF), osteogenic protein 1/bone morphogenetic protein 7 (OP-1/BMP-7), bone morphogenetic protein 2 and 12 (BMP-2, BMP-12), fibroblast growth factor 2/basic fibroblast growth factor (FGF-2/bFGF), and growth and differentiation factor 5 (GDF-5) have all been shown to act as anabolic stimuli on IVD cells. Moreover, in vitro anti-apoptotic effects of IGF-1, PDGF, and BMP-7 on human IVD cells have been described (Gruber et al., 2000; Wei et al., 2008). Similar to observations with chondrocytes, FGF-2 supplementation can also preserve the in vitro differentiation potential of NP cells (Tsai et al., 2007). An in vitro study on human IVD cells revealed that both recombinant human BMP-2 and -12 increased human NP cell matrix protein synthesis while having minimal effects on AF cells. However, adenoviral BMP-12 did increase matrix protein synthesis and proliferation in both NP and AF cells (Gilbertson et al., 2008). These responses to BMP-12, which is known to lack osteogenic activity in contrast to other factors of the BMP family, may give reason for further investigation. It was, however, reported that also BMP-2 did not show osteogenic effects in human cells from any IVD region (Kim et al., 2009b). BMP-2 exerted a mitogenic response on human AF cells and stimulated proteoglycan synthesis in NP cells, but it induced no significant expression of bone sialoprotein, DLX5, or osteocalcin mRNA.

GDF-5, also known as cartilage-derived morphogenetic protein 1 (CDMP-1) or BMP-14, has broadly been evaluated as anabolic factor for disc cells in vitro and in vivo. This growth factor stimulated proliferation and matrix production in bovine NP and to a lesser extent in AF cells (Chujo et al., 2006). After treatment of mouse NP cells with 10 and 100 ng/mL GDF-5 the DNA content, GAG production and collagen type II accumulation were increased, with more significant effects observed at the higher dose (Cui et al., 2008). In addition, treating the cells with GDF-5 protein increased the expression of the collagen type II and aggrecan genes in a dose-dependent manner, but decreased MMP 3 gene expression. Recently, CDMP 1 and 2 were identified in the human IVD, particularly in cells of the NP. A small decrease in the number of immunopositive cells was seen with degeneration. Treatment of human NP cells from degenerate IVD with CDMP showed an increase in aggrecan and type II collagen gene expression and increased production of proteoglycan (Le Maitre et al., 2009b).

As an autologous anabolic agent, platelet-rich plasma was effective in stimulating cell proliferation and extracellular matrix metabolism in porcine IVD cells cultured in alginate (Akeda et al., 2006). The response to platelet-rich plasma was greater in AF than NP cells.

Another autologous method for activation of NP cells is their coculture with mesenchymal stem cells. It was first demonstrated with rabbit cells that coculture systems with direct cell-to-cell contact between NP cells and bone marrow derived stromal cells notably activated proliferation and proteoglycan synthesis of the NP cells (Yamamoto et al., 2004). These positive results on the effect of coculture systems with direct cell-to-cell contact obtained in animal studies were recently also confirmed for human patients (Watanabe et al., 2010). Proliferation, DNA synthesis, and proteoglycan synthesis of NP cells isolated from surgical specimens were significantly enhanced after coculture with autologous bone marrow derived mesenchymal stem cells. Importantly, chromosome abnormalities and tumorigenesis, which have been associated with extensive cell stimulation, were not detected in the activated NP cells.

1.3.1.2 Disc Cell Transplantation Studies

Several studies have indicated that autologous chondrocyte transplantation is technically feasible and biologically appropriate to repair disc damage and retard disc degeneration. Restoration of the IVD by the use of autologous disc cell transplantation has been investigated in animal studies, and promising outcomes have made the initiation of clinical trials possible.

Reinsertion of autologous NP cells was studied in an experimental rabbit model of disc degeneration (Okuma et al., 2000). Coculture of the NP and AF cells stimulated proliferation of each cell type. Re-implantation of activated NP cells delayed the formation of clusters of chondrocyte-like cells and the destruction of the disc architecture. However, it needs to be emphasized that these results obtained with notochordal cells may not necessarily apply to mature central NP cells.

The feasibility of autologous disc cell implantation was demonstrated in the sand rat (psammomys obesus), a small rodent that spontaneously develops disc degeneration during aging (Gruber et al., 2002). Cells were harvested from a lumbar IVD, expanded in monolayer culture, labeled with agents that allow subsequent immunolocalization, and implanted in a second disc site of the donor animal. Immunocytological identification of engrafted cells showed that they integrated into the disc and were surrounded by normal matrix at time points up to 8 months after engraftment. Engrafted cells exhibited a spindle-shaped morphology in the annulus or a rounded chondrocyte-like morphology in the nucleus, indicating apposite integration into the tissue.

A canine study was performed to investigate the performance of autologous disc cell transplantation in a large animal model (Ganey et al., 2003). Lumbar disc tissue was harvested and cells expanded in culture with autologous serum for 12 weeks. Approximately 6 million cells were then re-implanted into the discs of same animals via percutaneous delivery. Discs were analysed after different time points up to 12 months. The implanted cells remained viable, maintained their proliferation ability and produced an extracellular matrix containing components similar to normal disc tissue. Proteoglycan and types I and II collagen were identified in the regenerated IVD matrix after cell transplantation. Additionally, a correlation between transplanting cells and retention of disc height was seen after longer intervals following transplantation. Importantly, there was no evidence of necrotic changes, vascularization, or bone in-growth.

In a rabbit disc degeneration model, transplantation of cells derived from a human NP cell line led to the preservation of the disc height and the NP matrix as evaluated by improved expression of the relevant matrix molecules (Iwashina et al., 2006). As transplantation of the human NP cell line was able to delay disc degeneration in this rabbit model, it was concluded that NP cell lines may become an alternative cell source for cell therapy of IVD degeneration.

The Euro Disc Randomized Trial was initiated to examine patients with both traumatic, less degenerative discs, and persistent symptoms (Hohaus et al., 2008; Meisel et al., 2007). Patients between 18 and 60 years undergoing surgical treatment for disc prolapse were enrolled in this study. Cells from sequestered disc material were expanded in culture, and at least 5 million cells were transplanted about 12 weeks following discectomy to assure the annulus had healed and would contain the cells. A simple, minimally-invasive technique was applied to support cell injection without further injury to the annulus. The first interim analysis revealed that patients who received autologous disc cell transplantation documented greater pain reduction after 2 years, compared to patients with no cell implantation following discectomy. Decreases in disc height over time were only observed in the group of patients who did not receive cell injections. Importantly, IVDs in segments adjacent to the discs that received the cell therapy also appeared to retain hydration when compared to those adjacent to levels that had undergone discectomy without cell intervention.

In spite of these promising results, a statement about the group of patients who will benefit most from autologous disc cell transplantation is not possible to date. Moreover, the technique is only applicable in patients who underwent discectomy. However, it would be preferable to initiate regenerative therapy in earlier stages of degeneration, when no matrix is lost from the disc yet. Harvesting of disc material from a healthy or early degenerative disc for the purpose of isolating cells for culture and re-implantation implies problems of donor site morbidity and risk of infection or nerve root injury. Accordingly, the medical basis for this intervention can hardly be approved. In addition, cells from degenerating discs may show an altered phenotype, implying increased expression of degrading enzymes, catabolic cytokines, and senescence markers (Le Maitre et al., 2004, 2005). It is evident that for prophylactic cell based disc regeneration, other cell types need to be available.

Given that the morphology and the phenotype of NP cells closely resemble the features of chondrocytes, the implantation of cartilage-derived cells into the IVD was considered. Survival and activity of autologous chondrocytes from elastic cartilage was studied in the lumbar IVD of rabbits (Gorensek et al., 2004). NP was evacuated and replaced with implants of auricular chondrocytes. Autologous cartilage implants were well tolerated by the host for up to 6 months, and hyaline-like cartilage but without any elastic fibers were identified in the NP. In an ex vivo study, rabbit knee articular chondrocytes transduced with adenovirus expressing human BMP-7 or human BMP-10 were injected into whole rabbit IVD explants (Zhang et al., 2008). Discs were maintained in culture for 1 to 2 months. The discs treated with chondrocytes/BMP-7 demonstrated a 50% increase in proteoglycan content within the NP compared to controls, while discs injected with chondrocytes/BMP-10 failed to show a significant increase in proteoglycan accumulation. This demonstrates the ability of transduced articular chondrocytes to survive and promote proteoglycan accumulation when transplanted into the IVD.

Although these results using chondrocytes from cartilage tissues appear promising, care should be taken in the interpretation of a tissue containing proteoglycans and type II collagen as functional NP tissue. There is no clear evidence that cell populations expressing the chondrocytic markers aggrecan and collagen type II are suitable for cell-based NP regeneration. It has been documented in several reports that the composition of the extracellular matrix, that is essential for the biomechanical function of the tissue, is not identical in NP and cartilage (Mwale et al., 2004; Vonk et al., 2010).

Given that the availability of suitable differentiated cells remains a major problem, research increasingly focuses on stem or progenitor cells as cell source for regenerative therapy. The general procedures for the use of cells for treatment of disc disease are summarized in Fig. 1.2.

1.3.2   STEM CELL BASED APPROACHES

1.3.2.1   In Vitro Differentiation of Mesenchymal Stem Cells

The cells that populate the NP have been shown to closely resemble the cells of cartilaginous tissues and have therefore been referred to as chondrocyte-like cells or disc chondrocytes. In particular, the expression of similar matrix molecules, the avascular nature, and the biomechanical function of cartilage and disc tissues result in related cell phenotypes, so that the phenotype of the NP cells has generally been termed "chondrocytic." Mesenchymal stem cells (MSCs) are progenitors that have the ability to differentiate along all connective tissue lineages including the chondrocytic profile (Pittenger et al., 1999). It has therefore been suggested that MSCs may also commit to the NP-like phenotype. MSCs represent an autologous supply of cells which can easily be harvested from different tissues, including bone marrow, adipose tissue, skeletal muscle, periosteum, and synovium, with bone marrow and adipose being the most widely investigated tissues. As with chondrocytic differentiation, various induction methods have been investigated to guide MSCs toward the expression of the disc cell phenotype. These include culture in a three-dimensional system, growth and differentiation factor supplementation, low oxygen tension, hydrostatic pressure as mechanical stimulus, co-culture with IVD cells, or a combination of these mediators.

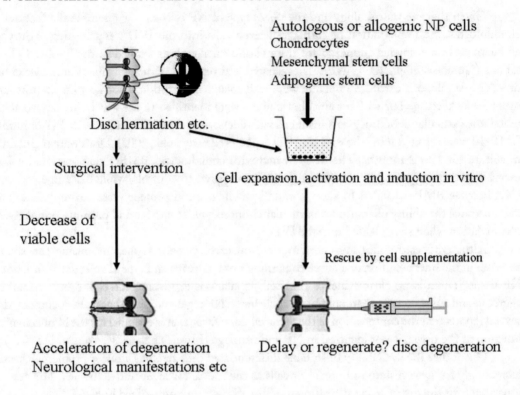

Figure 1.2: Therapeutic scenario for the use of cells for treatment of disc disease. The rescue of viable cells by supplementing various donor cells using in vitro culture techniques may help delay or regenerate the progressive degeneration process. (Source: Sakai, D., Future perspectives of cell-based therapy for intervertebral disc disease, *European Spine Journal* 17 Supp.4:452–458, 2008.)

Rat MSCs were cultured in alginate beads under hypoxic conditions (2% oxygen) with supplementation of TGF-beta (Risbud et al., 2004). Differentiation towards a phenotype consistent with that of NP cells was observed as demonstrated by the up-regulation of glucose transporter-3, MMP 2, collagen type II and type XI, and aggrecan mRNA and protein. In another study, human bone-marrow derived MSCs were maintained in spheroid culture under standard chondrogenic differentiation conditions in presence of TGF-beta, dexamethasone, and ascorbic acid (Steck et al., 2005). Interestingly, the induced cells within spheroids adopted a gene expression profile that more closely resembled native IVD cells than native articular chondrocytes. It should be noted, however, that this work did not distinguish NP and AF cells and that the data from IVD cells should be interpreted in favor of a gene expression profile closer to AF cells. Transfection with adenoviral Sox-9, a transcription factor involved in differentiation of MSCs along the chondrogenic lineage, in combination with culture in chondrogenic medium has also been used to induce an NP-like phenotype in monolayer or in three-dimensional scaffolds (Richardson et al., 2006a).

In spite of recent progress in the elucidation of the NP cell phenotype, there are still no unique molecular markers identified that can clearly distinguish articular chondrocytes from the "chondrocyte-like" cells found in the NP (Clouet et al., 2009; Lee et al., 2007; Minogue et al., 2010; Poiraudeau et al., 1999; Rutges et al., 2010; Sakai et al., 2009). For this reason, the majority of studies investigating the in vitro differentiation of MSCs towards the disc-like phenotype have used coculture systems of MSCs with cells derived from the IVD. In vitro coculture studies may also give indications about the behavior of MSCs implanted into a disc containing viable and active endogenous cells. Coculture with human NP cells demonstrated the ability to induce differentiation of human bone marrow derived MSCs as assessed by significant increases in Sox-9, type II collagen, and aggrecan gene expression and less pronounced increases in type I and type VI collagen (Richardson et al., 2006b). Direct cell-to-cell contact was necessary for MSC differentiation, and the extent of gene up-regulation was regulated by the cell ratio, with the optimum ratio found to be 75% NP cells and 25% MSCs. Similarly, when human NP cells were maintained in pellet coculture with human MSCs in different ratios, the 75%:25% and the 50%:50% ratios of NP cells to MSCs yielded the highest increases in extracellular matrix production (Sobajima et al., 2008). The mechanism of interaction between human NP cells and MSCs was investigated in three-dimensional alginate culture and in pellet culture (Vadala et al., 2008b). A decrease in collagen type I and an increase in collagen type II and aggrecan gene expression was observed in MSCs after culture with NP cells in alginate hydrogels that allowed short distance paracrine cell interactions. Further investigations by fluorescence in situ hybridization analysis revealed that cell fusion was not responsible for the MSC plasticity in the interaction with NP cells. To study the interaction of human NP cells with MSCs through longer distance paracrine stimulation, cell culture plates with inserts were used without direct cell-to-cell contact or exchange of cellular components (Yang et al., 2008). Whereas NP cell proliferation was remarkably increased by indirect coculture with the MSCs, the effect of NP cell derived soluble factors on MSCs was less prominent. Nevertheless, the suppression of collagen type I and up-regulation of collagen type II expression in MSCs indicated differentiation towards an NP-like phenotype, although high numbers of NP cells (80% NP cells versus 20% MSCs) were necessary to significantly induce type II collagen expression. It has to be considered that the cells were kept in monolayer culture, which hampers the chondrocytic or NP-like differentiation of MSCs.

Interactions between human NP cells and adipose tissue derived stem cells were studied in transwell coculture, employing both monolayer and micromass configurations (Lu et al., 2007). Notably, chondrogenic differentiation of the adipose stem cells was only achieved when both cell types were maintained in micromass culture. Differentiation was assessed by induction of collagen type II and aggrecan and down-regulation of osteopontin, collagen type I, and PPAR-γ. Since the cells were co-cultured without cell-to-cell contact, paracrine actions by soluble factors from micromass-cultured NP cells may have induced differentiation towards the NP-like phenotype. The direct culture environment of the adipose-derived stem cells also appears to play a role in that culture in a type II collagen hydrogel enhanced differentiation towards the chondrogenic or NP-like lineage as compared to culture in type I collagen gel (Lu et al., 2008).

Synovial tissue has been proposed to provide a superior type of MSCs for chondrogenesis as compared to other sources (Sakaguchi et al., 2005; Shirasawa et al., 2006). The in vitro differentiation potential of synovium-derived stem cells was therefore investigated in a coculture system with NP cells from a porcine source (Chen et al., 2009a). Coculture of equal amounts of synovium-derived stem cells and NP cells in a pellet system with cell-to-cell contact was performed in a low oxygen (5% O_2) environment with varying doses of TGF-beta supplementation. While other studies had reported that direct coculture with NP cells alone induced differentiation of MSCs, addition of TGF-beta was necessary to drive synovium-derived stem cell differentiation towards the NP-like phenotype. Specifically, the results suggested that treatment with TGF-beta at high concentration (30 ng/mL) yielded pellets from either NP cells or coculture showing an NP-like phenotype, whereas TGF-beta at lower concentrations (<10 ng/mL) resulted in pellets with a cartilage-like phenotype. This was concluded from the ratio of aggrecan to collagen type II that was markedly enhanced in NP and coculture pellets cultured with 30 ng/mL of TGF-beta. It had previously been reported that the ratio of glycosaminoglycan to hydroxyproline could be used to distinguish NP from cartilage tissue, in that a ratio of approximately 27 was found for the NP and a ratio of around 2 for cartilage (Mwale et al., 2004).

To examine the potential of muscle-derived stem cells, human NP cells were co-cultured with murine muscle-derived stem cells in monolayer culture in various ratios (Vadala et al., 2008a). The results demonstrated a synergistic effect of muscle-derived stem cells in culture with the NP cells, leading to increased proteoglycan synthesis and NP cell proliferation.

The effect of mechanical stimulation by application of hydrostatic pressure on rabbit bone marrow derived MSCs in indirect coculture with NP cells, both in separate three-dimensional alginate beads, was assessed in a transwell system (Kim et al., 2009a). Intermittent hydrostatic pressure (0.2 MPa; 2 min on/15 min off) was applied for 4 hours per day over 3 consecutive days. Confirming other investigations, only higher NP:MSC ratios, in this case 2:1, promoted the differentiation of the MSCs as evaluated by enhanced proteoglycan synthesis and diminished collagen type I gene expression. Moreover, differentiation activities were improved upon exposure to hydrostatic loading, which was shown by an increase in proteoglycans and slight elevation of aggrecan and Sox-9 gene expression, although the effect on gene transcription was minimal. From studies performed with human bone marrow derived MSCs it has been concluded that repetitive application of hydrostatic pressure over several days is necessary to achieve a significant effect. In addition, several studies have suggested that the peak effects may be delayed until some weeks after termination of loading, which hampers the interpretation of results and makes comparisons of different studies difficult (Elder and Athanasiou, 2009; Luo and Seedhom, 2007; Miyanishi et al., 2006).

Besides coculture with isolated NP cells, systems of culturing MSCs in presence of NP tissue were also described. Using a transwell system, adipose tissue derived rabbit mesenchymal stem cells embedded in alginate beads were exposed to samples of NP or AF tissue without direct contact (Li et al., 2005). Type II collagen and aggrecan gene expression levels were found to be 2-3 times higher in MSCs cultured with NP tissue than in MSCs in alginate alone or cultured with AF tissue. Of note, cell culture media were supplemented with 100 ng/mL GDF-5 and ascorbate, which also might have affected gene expression levels. When cultured with rodent IVD tissue, rat MSCs underwent morphological changes to form three-dimensional micromasses and expressed collagen type II, aggrecan, and Sox-9 at RNA and protein level after 14 days of coculture (Wei et al., 2009a). This indicates induction of the MSCs to express an NP-like phenotype under the direct influence of intact disc tissue.

The ultimate aim of such in vitro studies is to evaluate cell sources and to optimize conditions for MSC based NP regeneration strategies. There is, however, substantial evidence that the regenerative potential of MSCs is limited by the harsh milieu within the disc that may severely affect their survival. It is therefore fundamental to investigate the reaction of MSCs to the chemical environment of the IVD. Rat bone marrow derived MSCs were cultured in monolayer under glucose, osmolarity, and/or pH conditions approaching the atmosphere within a healthy or mildly degenerated disc (Wuertz et al., 2008). Low glucose levels (1 mg/mL) enhanced matrix biosynthesis and maintained cell proliferation, while high osmolarity (485 mOsm) and low pH conditions (pH 6.8) strongly reduced biosynthetic and proliferative activity of young and mature rat MSCs. It was concluded that the acidic conditions within degenerative discs most critically impair the survival and metabolic activity of implanted MSCs.

The survival and phenotype expression of human bone marrow derived MSCs following injection into bovine NP tissue were determined using an NP explant culture system (Le Maitre et al., 2009a). The injected MSCs survived well and appeared to differentiate towards an NP-like phenotype without any indication of mineralization until 4 weeks post injection. It is suggested that the NP tissue niche may provide an environment of native cells, matrix and growth factors that may have a role in the differentiation of the MSCs after implantation. However, this in vitro system could not fully replicate the hypoxic, nutrient restricted, acidic, and mechanically stressed in vivo surroundings, which needs to be considered in the interpretation of the results. Using an IVD organ culture system, the feasibility of MSC injection into cryopreserved bovine IVD was investigated (Chan et al., 2010). Although only a small proportion of injected MSCs survived, they showed some effect on the matrix protein gene expression. In another ex vivo culture system, chymopapain was used to partially digest NP tissue of porcine IVD with endplates, and the feasibility of injection of bone marrow derived MSCs and of platelet-rich plasma was evaluated (Chen et al., 2009b). Both the implantation of MSCs as well as application of PRP as a source of growth factors induced chondrogenic matrix production. Regenerative effects were confirmed in an in vivo porcine disc degeneration model. However, combined treatment with MSCs and PRP resulted in significantly osteogenic matrix accumulation rather than a synergistic or additive chondrogenic effect. This demonstrates that care should be taken when applying growth factors with osteogenic potential, in particular in combination with undifferentiated progenitor cells.

To eventually examine stem cell survival, differentiation, and activity within a living disc, in vivo studies that were performed using various small and large animal models are discussed in the following.

1.3.2.2 In Vivo Mesenchymal Stem Cell Transplantation Studies

There is still no clear evidence of the origin of the NP cells, and the phenotype of these cells is not yet sufficiently defined. This lack of basic knowledge may impede the development of MSC-based NP regeneration strategies. Nevertheless, various animal studies have demonstrated the feasibility of MSC transplantation into the IVD and have reported regenerative effects. Using a 15% hyaluronan gel as a carrier, fluorescence labeled MSCs were injected into rat coccygeal discs (Crevensten et al., 2004). Although the number of retained MSCs significantly decreased during the first two weeks after injection, a return to the initial cell number was noted after 4 weeks, while cell viability and disc height was maintained. This indicates that the injected cells started to proliferate within the rat disc.

Sakai et al. (Sakai et al., 2003, 2006) transplanted MSCs embedded in atelocollagen gel into discs of rabbits which had undergone NP aspiration to induce degeneration. After 24 weeks, discs of MSC-transplanted animals regained a disc height value of ~91% and MRI signal intensity of ~81% of normal controls, which were significantly higher compared to non-transplanted discs (Fig. 1.3). Furthermore, immunohistochemical and gene expression analysis confirmed enhanced proteoglycan accumulation in the MSC-transplanted discs. In another study, the differentiation of transplanted cells was specifically investigated (Sakai et al., 2005). Autologous MSCs, labeled with green fluorescent protein, were transplanted into mature rabbits. Green fluorescent protein positive cells were abundantly observed in the NP of cell-transplanted rabbit discs at 2 weeks and increased significantly by 48 weeks. Some MSCs also expressed cell-associated matrix molecules, such as type II collagen, keratan sulfate, chondroitin sulfate, aggrecan, and NP markers like hypoxia inducible factor 1 alpha, glutamine transporter 1, and matrix metalloproteinase 2 (Rajpurohit et al., 2002). This observation for the first time suggested that transplanted MSCs may have undergone site-dependent differentiation.

The potential of using allogenic bone marrow derived MSCs containing LacZ marker gene was shown in a rabbit model (Zhang et al., 2005). Follow-up data for up to 6 months indicated that the allogenic cells survived and were able to increase the proteoglycan and type II collagen content of the disc matrix, while type I collagen remained unchanged. In a similar study, allogenic bone marrow derived MSCs transduced with the LacZ marker gene were injected into the intact NP of rabbit lumbar discs, and the discs were analyzed up to 24 weeks after transplantation (Sobajima et al., 2008). MSCs were detected in histological sections of the discs without evidence of adverse reactions. Whereas at earlier time points, MSCs were localized within the NP and exhibited a rounded shape similar to NP cells, MSCs were detected in the transition zone and inner AF after 24 weeks, where they exhibited an altered more spindle shaped morphology similar to native AF cells. This suggests the possibility of stem cell migration, differentiation, and engraftment into the inner AF.

Normal control   Degeneration model   Stem cell transplanted

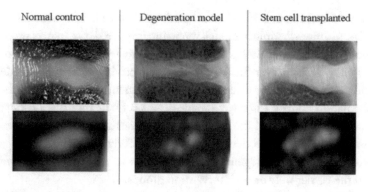

Figure 1.3: Macroscopic evaluation and T2 weighted magnetic resonance images (MRI) of undamaged IVD (normal control), IVD degeneration induced by nucleotomy (degeneration model), and mesenchymal stem cell transplantation (stem cell transplanted). The structure of the nucleus pulposus is better retained and MRI signal intensity has increased in the stem cell transplanted IVD. (Source: Sakai, D., Future perspectives of cell-based therapy for intervertebral disc disease, *European Spine Journal* 17 Supp.4:452–458, 2008.)

While these investigations in small animal models are useful for general practicability studies, their clinical relevance is limited due to the notable differences in the NP cell populations, NP size, and nutrition conditions between these species and human patients. A canine disc degeneration model was used to evaluate the efficacy of MSC transplantation in terms of preservation of the disc functionality and immune privilege (Hiyama et al., 2008). One million autologous MSCs were injected 4 weeks subsequent to NP aspiration in lumbar IVDs of mature beagle dogs (chondrodystrophoid breed), and animals were followed for another 8 weeks. Radiological, histological, biochemical, immunohistochemical, and gene expression analysis demonstrated clearly regenerative effects of the transplanted cells. Importantly, MSCs found in the NP 8 weeks after transplantation expressed Fas-ligand protein, which had not been detected in MSCs before transplantation. Fas-ligand, which is present in tissues with isolated immune privilege including the NP, plays an important role for NP maintenance. The expression of Fas-ligand indicates that MSC transplantation may also contribute to the preservation of the immune privilege of the disc environment.

Although adipose tissue derived cells have been investigated less extensively than bone marrow derived MSCs, recent work has indicated their potential for cell-based IVD regeneration (Hoogendoorn et al., 2008). Adipose tissue is considered as an abundant, expandable, and easily accessible source of MSCs. The use of these cells may thus eliminate the need for in vitro expansion, which raises the possibility of a one-step regenerative treatment method. Such a one-step procedure was evaluated in a goat model of disc degeneration, where chondroitinase ABC was used to induce disc damage (Hoogendoorn et al., 2008).

The efficacy of autologous adipose tissue derived stem cells in promoting disc regeneration was documented in a canine disc injury model, reporting improved disc matrix production and overall disc morphology (Ganey et al., 2009). Partial nucleotomy was performed at 3 lumbar levels and the animals were allowed to recover for 6 weeks before receiving either adipose derived stem cells in hyaluronic acid carrier, hyaluronic acid alone, or no treatment. Assessments of the 3 experimental discs plus the 2 adjacent control discs were made after up to 12 months. Discs receiving adipose derived cells more closely resembled the healthy controls as evidenced in matrix translucency, compartmentalization of the AF, and cell density within the NP. Matrix analysis for type II collagen and aggrecan demonstrated a superior regenerative stimulation in the discs treated with adipose stem cells compared to the carrier only or no intervention groups.

Recent studies investigated the fate of transplanted human MSCs into IVDs in xenogeneic animal models. Human bone marrow derived stem cells were separated in hematopoietic ($CD34^+$) and non-hematopoietic MSCs including ($CD34^-$) fractions, fluorescent-labeled, and injected into rat coccygeal discs (Wei et al., 2009b). $CD34^-$ cells were detected in the NP of 67% of the discs until day 42 after injection. By day 21, these cells differentiated towards a chondrocytic phenotype as shown by expression of collagen type II and Sox-9. In contrast, $CD34^+$ cells decreased significantly by day 10 and were undetectable in the discs at day 21. Neither inflammatory nor immune cells infiltrated the NP, in spite of the xenogeneic nature of the implanted cells, confirming the immune privileged characteristics of the NP.

The survival and function of human MSCs after transplantation into injured porcine discs was also studied (Henriksson et al., 2009b). Nucleotomy was performed in lumbar discs of minipigs. After 2 weeks, 5×10^5 human bone marrow derived MSCs were injected with or without a hydrogel carrier. The animals were followed for up to 6 months. Human MSCs were detected in ~90% of the injected discs. Immunostaining for aggrecan and collagen type II was observed in transplanted MSCs after 3 and 6 months, suggesting differentiation towards disc-like cells. Gene expression of collagen IIA, collagen IIB, versican, collagen I, aggrecan, and Sox-9 were noticed in injected discs at 3 and 6 months. Moreover, the combination with a three-dimensional hydrogel carrier seemed to facilitate differentiation and survival of MSCs in the disc. These promising results render transplantation of human stem cells into mature animal discs a valuable in vivo model to study the capacity of MSCs and their subpopulations for NP regeneration and to optimize the procedures.

The behavior of murine embryonic stem cell derived chondroprogenitors implanted into a rabbit disc was evaluated in a percutaneous puncture disc degeneration model (Sheikh et al., 2009). Eight weeks after injection, new notochordal cell populations were seen in the degenerated discs injected with the embryonic stem cell derivatives. Again neither inflammatory nor immune responses were observed. The possible advantage of using embryonic stem cells rather than adult cell populations for NP regenerative therapy needs to be investigated. In view of the potential significance of notochordal cells for the maintenance of the NP homeostasis, the use of more primitive stem cells warrants consideration. However, ethical concerns severely limit the use of embryonic cell sources. Interestingly, an increase in endogenous notochordal cells was also found after injection of bone marrow derived MSCs in murine IVDs degenerated by annular puncture (Yang et al., 2009). The study confirmed regenerative effects of injected MSCs observed in other animal models. In particular, the authors highlight that two mechanisms are likely to be responsible for the disc degeneration arrest provided by implanted stem cells: Both (i) differentiation of the stem cells under the influence of the NP environment and (ii) anabolic stimulation of the endogenous cells appear to play a role.

1.4   LIMITATIONS AND FUTURE PERSPECTIVE

New imaging techniques and recent advances in cell biology and genetics hold promise for better understanding of the disc physiology and pathophysiology, earlier and more specific diagnosis, and targeted treatment. New treatment should aim at preventing adjacent tissue degeneration, abnormal load and load distribution, and eventually spinal stenosis. Restoring or at least maintaining disc height is a primary requirement to achieve this ambition. Although clinical trials using autologous disc cell transplantation are already ongoing and in vitro and animal studies have demonstrated the potential of MSC injection, there are still a number of issues to solve before cell therapy can be considered as a widely accepted treatment for NP regeneration.

It needs to be borne in mind that the functions of the IVD require a mechanically stable structure with a highly specialized matrix to confer flexibility and physical strength to the spine. Autologous NP cells would be the preferred cell population since they are supposed to be adapted to the disc environment and synthesize the correct matrix components. However, unlike chondrocytes that can be obtained from non-weight bearing areas of intact cartilage and used for autologous chondrocyte transplantation for cartilage repair, there is no source of autologous disc cells from healthy regions. Harvesting an IVD biopsy will inevitably provoke degeneration of the respective disc and is therefore not feasible. On the other hand, autologous disc cells isolated from herniated or degenerate disc tissue removed during routine surgery may express an altered phenotype. Signs of senescence have regularly been observed in cells from aged or diseased discs. Senescent cells are viable, but cannot proliferate and show alterations in nuclear structure, gene expression, protein processing, and cell metabolism. Cells from surgical specimens also often show evidence of a catabolic and/or pro-inflammatory status and are thus not ideal candidates for regenerative treatment. Furthermore, it can be difficult to separate annulus from nucleus tissue, resulting in mixed undefined disc cell populations. Human disc tissue is characterized by a low cell density, and the number of viable cells may be severely decreased in affected discs. In vitro expansion will therefore be required to obtain sufficient amounts of viable cells for implantation. When NP cells are cultured in monolayer, they tend to rapidly dedifferentiate and express a phenotype that deviates from the NP profile. This may alter the cells' behavior upon re-implantation into the disc and may result in inappropriate matrix production. It is also conceivable that the cells will re-differentiate when they are placed back in the IVD environment. Thus, it would be of importance to monitor the phenotype of the disc cells directly after isolation, after in vitro culture, and after re-implantation. Although the latter is not feasible in clinical practice, it would be desirable to rely on a standard phenotypic profile to assess the quality of a cell population before implantation.

Efforts have been made to improve the proliferation and metabolic activity of isolated disc cells before re-implantation. A promising method is the pre-conditioning of NP cells isolated from surgical samples in coculture with autologous bone marrow derived MSCs (Watanabe et al., 2010). There are concerns that with extensive activation and population doublings cells can become senescent or the cell karyotype can change and the risk for tumor formation can rise. Neither chromosome changes nor signs of tumor formation were noted in human NP cells after coculture with MSCs, indicating that this method for NP cell activation is safe in this respect. Other ways to accelerate the proliferation and metabolism of NP cells in vitro may include supplementation with growth factors or autologous platelet rich plasma or platelet lysate. However, similar concerns as addressed above arise. It needs to be noted that in vitro culture conditions generally differ from the in vivo environment of NP cells. Thus, the in vitro culture time, the required manipulations, and the contact with exogenous factors need to be kept at a minimum in favor of patient safety.

Injection of MSCs appears to be an attractive alternative to NP cell transplantation. MSCs can be obtained relatively easily, and their hypo-immunogenic and immunosuppressive properties in combination with the relatively immune-privileged nature of the NP make allogenic transplantation a valuable option (Ryan et al., 2005). In fact, a recent case study reports on the successful application of autologous MSCs in two patients with marked lumbar IVD degeneration (Yoshikawa et al., 2010). However, several concerns need to be clarified before MSC transplantation can be considered for broad clinical application. Although in vitro and in vivo experiments indicate that MSCs can differentiate toward the "discogenic" lineage, no valid method for obtaining disc-like cells from MSCs has been identified so far. In addition, the quality of the matrix produced by induced cells that show an "NP-like" expression profile is uncertain and requires further biomechanical evaluation. Most importantly, there is still no clear marker available that can define NP cells and distinguish them from chondrocytes found in articular cartilage. Recent work has addressed this issue applying large scale expression profiling and comparing disc and cartilage cells (Clouet et al., 2009; Lee et al., 2007; Minogue et al., 2010; Rutges et al., 2010; Sakai et al., 2009). Findings from animal species, however, can only be translated to human cells after careful evaluation, given that fundamental differences were observed in gene and protein expression patterns between rodent, large animal and human cells. Age- and degeneration-related divergences and evidence of the presence of more than one cell type in the NP further complicate the definition of an NP cell phenotype. Ultimately, patient safety remains a major concern also for MSC-based approaches with potential complications related to phenotypic transformations and tumor formation.

While animal models might serve to investigate basic mechanisms of disc homeostasis, they are often not valid for evaluation of new therapeutic approaches (Alini et al., 2008). It is therefore questionable whether encouraging animal studies will translate into successful clinical application. First, the development, nutrition, and mechanical function of the IVD substantially differ between species. Second, animal models for disc degeneration may not replicate the events occurring in human disc disease. Provided that the processes of disc degeneration are not consistent, it is very difficult, or almost not possible, to establish an animal model for natural progression of disc disease.

In view of this fact, every model should be clearly matched to the study objective in order to avoid misinterpretations.

On the other hand, investigations on isolated human cells in culture may not provide relevant information since it is almost impossible to replicate in a cell culture system the biochemical, biophysical, and biomechanical conditions that are experienced in the disc tissue. In particular, the rate of matrix synthesis may be substantially delayed in the intradiscal environment. Whole IVD organ culture systems offer a valuable opportunity to test future treatment approaches in a disc environment ex vivo (Illien-Junger et al., 2010; Junger et al., 2009). In particular, large animal discs that approximate human discs can be assessed without the need of costly and sometimes ethically questionable animal studies. Such organ culture models should be used for feasibility studies under relevant conditions, while the efficacy must ultimately be determined in a large animal model.

There is still limited knowledge about the survival of cells transplanted to the NP and their phenotypical stability over longer time periods in vivo. It is obvious that enormous demands are made on implanted cells; thereby, IVD nutrition seems to be one of the main challenges. It is widely accepted that nutritional deprivation is an age-related process and is associated with degenerative changes. Nevertheless, the explicit contribution of inadequate nutrient supply to the progression of disc degeneration is not without controversy. Attempts need to be made to reliably determine the nutrition state of affected discs, and concomitantly methods to improve access to nutrients need to be found. The differentiation status of the native disc cells and their homeostasis will have an impact of the therapeutic success of any cell therapy. In addition, the optimal cell number to be applied in a therapeutic setting needs to be carefully evaluated, as increasing the number of cells will require higher amounts of nutrients.

Different regenerative strategies will be preferred according to the conditions, age of the patient, and degenerative stage. Since the existence of endogenous, provoking factors may render any treatment approach ineffective or only temporary. Furthermore, targeting proinflammatory signaling pathways might be successful in certain situations. In this context, clear criteria need to be defined for selection of the appropriate patients, in order to achieve best efficacy of a specific therapy.

In conclusion, the limited number of viable cells in degenerate discs is a major concern for the application of therapeutic molecules or genes, since stimulation of the remaining cells may be insufficient to restore the disc matrix. Supplementing the cells of the disc by adding new cells either alone or in combination with an appropriate scaffold may overcome this problem. There is increasing agreement that not only the NP but also the AF and eventually the cartilaginous endplate should be considered as therapeutic targets for effective cell-based therapy. Finally, in situ therapy approaches using chemotactic molecules to recruit endogenous progenitor cells appear to be notably more attractive, avoiding all the technical and biological problems associated with the exogenous delivery of cells. Recent work indicates that progenitor cell populations are present in degenerate disc tissue and in defined areas (niches) of healthy discs (Henriksson et al., 2009a; Risbud et al., 2007). Activation of endogenous regeneration mechanisms by targeted mobilization of progenitor cells would be the ultimate goal; further fundamental studies will elucidate the achievability of this ambition.

1.5   ACKNOWLEDGEMENTS

The authors would like to acknowledge the Swiss National Science Foundation for their support under the research grant No. 320030-116818.

REFERENCES

Adams,MA, P J Roughley, (2006), What is intervertebral disc degeneration, and what causes it?: Spine (Phila Pa 1976.), v. 31, p. 2151–2161. DOI: 10.1097/01.brs.0000231761.73859.2c 4, 6

Aguiar,DJ, S L Johnson, T R Oegema, (1999), Notochordal cells interact with nucleus pulposus cells: regulation of proteoglycan synthesis: Exp.Cell Res., v. 246, p. 129–137. DOI: 10.1006/excr.1998.4287 7

Akeda,K, H S An, R Pichika, M Attawia, E J Thonar, M E Lenz, A Uchida, K Masuda, (2006), Platelet-rich plasma (PRP) stimulates the extracellular matrix metabolism of porcine nucleus pulposus and anulus fibrosus cells cultured in alginate beads: Spine (Phila Pa 1976.), v. 31, p. 959–966. DOI: 10.1097/01.brs.0000214942.78119.24 11

Alini,M, S M Eisenstein, K Ito, C Little, A A Kettler, K Masuda, J Melrose, J Ralphs, I Stokes, H J Wilke, (2008), Are animal models useful for studying human disc disorders/degeneration?: Eur.Spine J., v. 17, p. 2–19. DOI: 10.1007/s00586-007-0414-y 22

Antoniou,J, T Steffen, F Nelson, N Winterbottom, A P Hollander, R A Poole, M Aebi, M Alini, (1996), The human lumbar intervertebral disc: evidence for changes in the biosynthesis and denaturation of the extracellular matrix with growth, maturation, ageing, and degeneration: J.Clin.Invest, v. 98, p. 996–1003. DOI: 10.1172/JCI118884 5

Battie,MC, T Videman, (2006), Lumbar disc degeneration: epidemiology and genetics: J.Bone Joint Surg.Am., v. 88 Suppl 2, p. 3–9. DOI: 10.2106/JBJS.E.01313 6

Battie,MC, T Videman, L E Gibbons, L D Fisher, H Manninen, K Gill, (1995), 1995 Volvo Award in clinical sciences. Determinants of lumbar disc degeneration. A study relating lifetime exposures and magnetic resonance imaging findings in identical twins: Spine (Phila Pa 1976.), v. 20, p. 2601–2612. 6

Benneker,LM, P F Heini, M Alini, S E Anderson, K Ito, (2005), 2004 Young Investigator Award Winner: vertebral endplate marrow contact channel occlusions and intervertebral disc degeneration: Spine (Phila Pa 1976.), v. 30, p. 167–173. DOI: 10.1097/01.brs.0000150833.93248.09 5

Bibby,SR, J C Fairbank, M R Urban, J P Urban, (2002), Cell viability in scoliotic discs in relation to disc deformity and nutrient levels: Spine (Phila Pa 1976.), v. 27, p. 2220–2228. 8

Bibby,SR, D A Jones, R M Ripley, J P Urban, (2005), Metabolism of the intervertebral disc: effects of low levels of oxygen, glucose, and pH on rates of energy metabolism of bovine nucleus pulposus cells: Spine (Phila Pa 1976.), v. 30, p. 487–496. DOI: 10.1097/01.brs.0000154619.38122.47 8

Bibby,SR, J P Urban, (2004), Effect of nutrient deprivation on the viability of intervertebral disc cells: Eur.Spine J., v. 13, p. 695–701. DOI: 10.1007/s00586-003-0616-x 9

Blumenthal,S, P C McAfee, R D Guyer, S H Hochschuler, F H Geisler, R T Holt, R Garcia, Jr., J J Regan, D D Ohnmeiss, (2005), A prospective, randomized, multicenter Food and Drug Administration investigational device exemptions study of lumbar total disc replacement with the CHARITE artificial disc versus lumbar fusion: part I: evaluation of clinical outcomes: Spine (Phila Pa 1976.), v. 30, p. 1565–1575. DOI: 10.1097/01.brs.0000170561.25636.1c 2

Bono,CM, C K Lee, (2004), Critical analysis of trends in fusion for degenerative disc disease over the past 20 years: influence of technique on fusion rate and clinical outcome: Spine (Phila Pa 1976.), v. 29, p. 455–463. 2

Boos,N, A G Nerlich, I Wiest, M K von der, M Aebi, (1997), Immunolocalization of type X collagen in human lumbar intervertebral discs during ageing and degeneration: Histochem.Cell Biol., v. 108, p. 471–480. DOI: 10.1007/s004180050187 5

Boos,N, S Weissbach, H Rohrbach, C Weiler, K F Spratt, A G Nerlich, (2002), Classification of age-related changes in lumbar intervertebral discs: 2002 Volvo Award in basic science: Spine (Phila Pa 1976.), v. 27, p. 2631–2644. 1, 7

Brox,JI, O Reikeras, O Nygaard, R Sorensen, A Indahl, I Holm, A Keller, T Ingebrigtsen, O Grundnes, J E Lange, A Friis, (2006), Lumbar instrumented fusion compared with cognitive intervention and exercises in patients with chronic back pain after previous surgery for disc herniation: a prospective randomized controlled study: Pain, v. 122, p. 145–155. DOI: 10.1016/j.pain.2006.01.027 2

Butler,D, J H Trafimow, G B Andersson, T W McNeill, M S Huckman, (1990), Discs degenerate before facets: Spine (Phila Pa 1976.), v. 15, p. 111–113. 5

Cappello,R, J L Bird, D Pfeiffer, M T Bayliss, J Dudhia, (2006), Notochordal cell produce and assemble extracellular matrix in a distinct manner, which may be responsible for the maintenance of healthy nucleus pulposus: Spine (Phila Pa 1976.), v. 31, p. 873–882. DOI: 10.1097/01.brs.0000209302.00820.fd 7

Chan,SC, B Gantenbein-Ritter, V Y Leung, D Chan, K M Cheung, K Ito, (2010), Cryopreserved intervertebral disc with injected bone marrow-derived stromal cells: a feasibility study using organ culture: Spine J, v. 10, p. 486–496. DOI: 10.1016/j.spinee.2009.12.019 17

Chen,B, J Fellenberg, H Wang, C Carstens, W Richter, (2005), Occurrence and regional distribution of apoptosis in scoliotic discs: Spine (Phila Pa 1976.), v. 30, p. 519–524. DOI: 10.1097/01.brs.0000154652.96975.1f 8

Chen,S, S E Emery, M Pei, (2009)a, Coculture of synovium-derived stem cells and nucleus pulposus cells in serum-free defined medium with supplementation of transforming growth factor-beta1: a potential application of tissue-specific stem cells in disc regeneration: Spine (Phila Pa 1976.), v. 34, p. 1272–1280. DOI: 10.1097/BRS.0b013e3181a2b347 16

Chen,WH, H Y Liu, W C Lo, S C Wu, C H Chi, H Y Chang, S H Hsiao, C H Wu, W T Chiu, B J Chen, W P Deng, (2009)b, Intervertebral disc regeneration in an ex vivo culture system using mesenchymal stem cells and platelet-rich plasma: Biomaterials, v. 30, p. 5523–5533. DOI: 10.1016/j.biomaterials.2009.07.019 17

Chujo,T, H S An, K Akeda, K Miyamoto, C Muehleman, M Attawia, G Andersson, K Masuda, (2006), Effects of growth differentiation factor-5 on the intervertebral disc–in vitro bovine study and in vivo rabbit disc degeneration model study: Spine (Phila Pa 1976.), v. 31, p. 2909–2917. DOI: 10.1097/01.brs.0000248428.22823.86 10

Clouet,J, G Grimandi, M Pot-Vaucel, M Masson, H B Fellah, L Guigand, Y Cherel, E Bord, F Rannou, P Weiss, J Guicheux, C Vinatier, (2009), Identification of phenotypic discriminating markers for intervertebral disc cells and articular chondrocytes: Rheumatology.(Oxford), v. 48, p. 1447–1450. DOI: 10.1093/rheumatology/kep262 15, 22

Crevensten,G, A J Walsh, D Ananthakrishnan, P Page, G M Wahba, J C Lotz, S Berven, (2004), Intervertebral disc cell therapy for regeneration: mesenchymal stem cell implantation in rat intervertebral discs: Ann.Biomed.Eng, v. 32, p. 430–434. DOI: 10.1023/B:ABME.0000017545.84833.7c 18

Cui,M, Y Wan, D G Anderson, F H Shen, B M Leo, C T Laurencin, G Balian, X Li, (2008), Mouse growth and differentiation factor-5 protein and DNA therapy potentiates intervertebral disc cell aggregation and chondrogenic gene expression: Spine J., v. 8, p. 287–295. DOI: 10.1016/j.spinee.2007.05.012 11

Deyo,RA, S K Mirza, (2006), Trends and variations in the use of spine surgery: Clin.Orthop.Relat Res., v. 443, p. 139–146. DOI: 10.1097/01.blo.0000198726.62514.75 2

Di Martino,A, A R Vaccaro, J Y Lee, V Denaro, M R Lim, (2005), Nucleus pulposus replacement: basic science and indications for clinical use: Spine (Phila Pa 1976.), v. 30, p. S16-S22. DOI: 10.1097/01.brs.0000174530.88585.32 2

Duance,VC, J K Crean, T J Sims, N Avery, S Smith, J Menage, S M Eisenstein, S Roberts, (1998), Changes in collagen cross-linking in degenerative disc disease and scoliosis: Spine (Phila Pa 1976.), v. 23, p. 2545–2551. 4

Eck,JC, S C Humphreys, S D Hodges, (1999), Adjacent-segment degeneration after lumbar fusion: a review of clinical, biomechanical, and radiologic studies: Am.J.Orthop.(Belle.Mead NJ), v. 28, p. 336–340. 6

Elder,BD, K A Athanasiou, (2009), Hydrostatic pressure in articular cartilage tissue engineering: from chondrocytes to tissue regeneration: Tissue Eng Part B Rev., v. 15, p. 43–53. DOI: 10.1089/ten.teb.2008.0435 16

Erwin,WM, R D Inman, (2006), Notochord cells regulate intervertebral disc chondrocyte proteoglycan production and cell proliferation: Spine (Phila Pa 1976.), v. 31, p. 1094–1099. DOI: 10.1097/01.brs.0000216593.97157.dd 7

Fairbank,J, H Frost, J Wilson-MacDonald, L M Yu, K Barker, R Collins, (2005), Randomised controlled trial to compare surgical stabilisation of the lumbar spine with an intensive rehabilitation programme for patients with chronic low back pain: the MRC spine stabilisation trial: BMJ, v. 330, p. 1233. DOI: 10.1136/bmj.38441.620417.8F 2

Freemont,AJ, T E Peacock, P Goupille, J A Hoyland, J O'Brien, M I Jayson, (1997), Nerve ingrowth into diseased intervertebral disc in chronic back pain: Lancet, v. 350, p. 178–181. DOI: 10.1016/S0140-6736(97)02135-1 5

Frymoyer,JW, (1992), Predicting disability from low back pain: Clin.Orthop.Relat Res., p. 101–109. 1

Ganey,T, W C Hutton, T Moseley, M Hedrick, H J Meisel, (2009), Intervertebral disc repair using adipose tissue-derived stem and regenerative cells: experiments in a canine model: Spine (Phila Pa 1976.), v. 34, p. 2297–2304. DOI: 10.1097/BRS.0b013e3181a54157 20

Ganey,T, J Libera, V Moos, O Alasevic, K G Fritsch, H J Meisel, W C Hutton, (2003), Disc chondrocyte transplantation in a canine model: a treatment for degenerated or damaged intervertebral disc: Spine (Phila Pa 1976.), v. 28, p. 2609–2620. 12

Gibson,JN, G Waddell, (2005), Surgery for degenerative lumbar spondylosis: updated Cochrane Review: Spine (Phila Pa 1976.), v. 30, p. 2312–2320. DOI: 10.1097/01.brs.0000182315.88558.9c 2

Gilbertson,L, S H Ahn, P N Teng, R K Studer, C Niyibizi, J D Kang, (2008), The effects of recombinant human bone morphogenetic protein-2, recombinant human bone morphogenetic protein-12, and adenoviral bone morphogenetic protein-12 on matrix synthesis in human annulus fibrosis and nucleus pulposus cells: Spine J., v. 8, p. 449–456. DOI: 10.1016/j.spinee.2006.11.006 10

Gorensek,M, C Jaksimovic, N Kregar-Velikonja, M Gorensek, M Knezevic, M Jeras, V Pavlovcic, A Cor, (2004), Nucleus pulposus repair with cultured autologous elastic cartilage derived chondrocytes: Cell Mol.Biol.Lett., v. 9, p. 363–373. 13

Gray,DT, R A Deyo, W Kreuter, S K Mirza, P J Heagerty, B A Comstock, L Chan, (2006), Population-based trends in volumes and rates of ambulatory lumbar spine surgery: Spine (Phila Pa 1976.), v. 31, p. 1957–1963. DOI: 10.1097/01.brs.0000229148.63418.c1 2

Gruber,HE, E N Hanley, Jr., (1998), Analysis of aging and degeneration of the human intervertebral disc. Comparison of surgical specimens with normal controls: Spine (Phila Pa 1976.), v. 23, p. 751–757. 7

Gruber,HE, J A Ingram, H J Norton, E N Hanley, Jr., (2007), Senescence in cells of the aging and degenerating intervertebral disc: immunolocalization of senescence-associated beta-galactosidase in human and sand rat discs: Spine (Phila Pa 1976.), v. 32, p. 321–327. DOI: 10.1097/01.brs.0000253960.57051.de 8

Gruber,HE, T L Johnson, K Leslie, J A Ingram, D Martin, G Hoelscher, D Banks, L Phieffer, G Coldham, E N Hanley, Jr., (2002), Autologous intervertebral disc cell implantation: a model using Psammomys obesus, the sand rat: Spine (Phila Pa 1976.), v. 27, p. 1626–1633. 11

Gruber,HE, H J Norton, E N Hanley, Jr., (2000), Anti-apoptotic effects of IGF-1 and PDGF on human intervertebral disc cells in vitro: Spine (Phila Pa 1976.), v. 25, p. 2153–2157. 10

Grunhagen,T, G Wilde, D M Soukane, S A Shirazi-Adl, J P Urban, (2006), Nutrient supply and intervertebral disc metabolism: J.Bone Joint Surg.Am., v. 88 Suppl 2, p. 30–35. DOI: 10.2106/JBJS.E.01290 4

Henriksson,H, M Thornemo, C Karlsson, O Hagg, K Junevik, A Lindahl, H Brisby, (2009)a, Identification of cell proliferation zones, progenitor cells and a potential stem cell niche in the intervertebral disc region: a study in four species: Spine (Phila Pa 1976.), v. 34, p. 2278–2287. DOI: 10.1097/BRS.0b013e3181a95ad2 23

Henriksson,HB, T Svanvik, M Jonsson, M Hagman, M Horn, A Lindahl, H Brisby, (2009)b, Transplantation of human mesenchymal stems cells into intervertebral discs in a xenogeneic porcine model: Spine (Phila Pa 1976.), v. 34, p. 141–148. DOI: 10.1097/BRS.0b013e31818f8c20 20

Hiyama,A, J Mochida, T Iwashina, H Omi, T Watanabe, K Serigano, F Tamura, D Sakai, (2008), Transplantation of mesenchymal stem cells in a canine disc degeneration model: J.Orthop.Res., v. 26, p. 589–600. DOI: 10.1002/jor.20584 19

Hohaus,C, T M Ganey, Y Minkus, H J Meisel, (2008), Cell transplantation in lumbar spine disc degeneration disease: Eur.Spine J., v. 17 Suppl 4, p. 492–503. DOI: 10.1007/s00586-008-0750-6 12

Holm,S, A Maroudas, J P Urban, G Selstam, A Nachemson, (1981), Nutrition of the intervertebral disc: solute transport and metabolism: Connect.Tissue Res., v. 8, p. 101–119. DOI: 10.3109/03008208109152130 5

Hoogendoorn,RJ, Z F Lu, R J Kroeze, R A Bank, P I Wuisman, M N Helder, (2008), Adipose stem cells for intervertebral disc regeneration: current status and concepts for the future: J.Cell Mol.Med., v. 12, p. 2205–2216. DOI: 10.1111/j.1582-4934.2008.00291.x 19

Horner,HA, S Roberts, R C Bielby, J Menage, H Evans, J P Urban, (2002), Cells from different regions of the intervertebral disc: effect of culture system on matrix expression and cell phenotype: Spine (Phila Pa 1976.), v. 27, p. 1018–1028. 9

Horner,HA, J P Urban, (2001), 2001 Volvo Award Winner in Basic Science Studies: Effect of nutrient supply on the viability of cells from the nucleus pulposus of the intervertebral disc: Spine (Phila Pa 1976.), v. 26, p. 2543–2549. 8

Hutton,WC, W A Elmer, S D Boden, S Hyon, Y Toribatake, K Tomita, G A Hair, (1999), The effect of hydrostatic pressure on intervertebral disc metabolism: Spine (Phila Pa 1976.), v. 24, p. 1507–1515. 9

Iatridis,JC, J J Maclean, P J Roughley, M Alini, (2006), Effects of mechanical loading on intervertebral disc metabolism in vivo: J.Bone Joint Surg.Am., v. 88 Suppl 2, p. 41–46. DOI: 10.2106/JBJS.E.01407 6

Illien-Junger,S, B Gantenbein-Ritter, S Grad, P Lezuo, S J Ferguson, M Alini, K Ito, (2010), The Combined Effects of Limited Nutrition and High-Frequency Loading on Intervertebral Discs With Endplates: Spine (Phila Pa 1976.). DOI: 10.1097/BRS.0b013e3181c48019 6, 23

Inkinen,RI, M J Lammi, S Lehmonen, K Puustjarvi, E Kaapa, M I Tammi, (1998), Relative increase of biglycan and decorin and altered chondroitin sulfate epitopes in the degenerating human intervertebral disc: J.Rheumatol., v. 25, p. 506–514. 4

Ishihara,H, D S McNally, J P Urban, A C Hall, (1996), Effects of hydrostatic pressure on matrix synthesis in different regions of the intervertebral disk: J.Appl.Physiol, v. 80, p. 839–846. 10

Iwashina,T, J Mochida, D Sakai, Y Yamamoto, T Miyazaki, K Ando, T Hotta, (2006), Feasibility of using a human nucleus pulposus cell line as a cell source in cell transplantation therapy for intervertebral disc degeneration: Spine (Phila Pa 1976.), v. 31, p. 1177–1186. DOI: 10.1097/01.brs.0000217687.36874.c4 12

Johnson,WE, B Caterson, S M Eisenstein, D L Hynds, D M Snow, S Roberts, (2002), Human intervertebral disc aggrecan inhibits nerve growth in vitro: Arthritis Rheum., v. 46, p. 2658–2664. DOI: 10.1002/art.10585 5

Johnson,WE, S M Eisenstein, S Roberts, (2001), Cell cluster formation in degenerate lumbar intervertebral discs is associated with increased disc cell proliferation: Connect.Tissue Res., v. 42, p. 197–207. DOI: 10.3109/03008200109005650 8

Johnson,WE, S Roberts, (2007), 'Rumours of my death may have been greatly exaggerated': a brief review of cell death in human intervertebral disc disease and implications for cell transplantation therapy: Biochem.Soc.Trans., v. 35, p. 680–682. DOI: 10.1042/BST0350680 8

Jones,P, L Gardner, J Menage, G T Williams, S Roberts, (2008), Intervertebral disc cells as competent phagocytes in vitro: implications for cell death in disc degeneration: Arthritis Res.Ther., v. 10, p. R86. DOI: 10.1186/ar2466 8

Junger,S, B Gantenbein-Ritter, P Lezuo, M Alini, S J Ferguson, K Ito, (2009), Effect of limited nutrition on in situ intervertebral disc cells under simulated-physiological loading: Spine (Phila Pa 1976.), v. 34, p. 1264–1271. DOI: 10.1097/BRS.0b013e3181a0193d 9, 23

Kalichman,L, D J Hunter, (2008), The genetics of intervertebral disc degeneration. Associated genes: Joint Bone Spine, v. 75, p. 388–396. DOI: 10.1016/j.jbspin.2007.11.002 6

Kang,JD, M Stefanovic-Racic, L A McIntyre, H I Georgescu, C H Evans, (1997), Toward a bio-chemical understanding of human intervertebral disc degeneration and herniation. Contributions of nitric oxide, interleukins, prostaglandin E2, and matrix metalloproteinases: Spine (Phila Pa 1976.), v. 22, p. 1065–1073. 4

Kasra,M, V Goel, J Martin, S T Wang, W Choi, J Buckwalter, (2003), Effect of dynamic hydrostatic pressure on rabbit intervertebral disc cells: J.Orthop.Res., v. 21, p. 597–603. DOI: 10.1016/S0736-0266(03)00027-5 9

Kasra,M, W D Merryman, K N Loveless, V K Goel, J D Martin, J A Buckwalter, (2006), Frequency response of pig intervertebral disc cells subjected to dynamic hydrostatic pressure: J.Orthop.Res., v. 24, p. 1967–1973. DOI: 10.1002/jor.20253 9

Khan,SN, A J Stirling, (2007), Controversial topics in surgery: degenerative disc disease: disc re-placement. Against: Ann.R.Coll.Surg.Engl., v. 89, p. 6–11. DOI: 10.1308/003588407X160792 2

Kim,DH, S H Kim, S J Heo, J W Shin, S W Lee, S A Park, J W Shin, (2009)a, Enhanced differentiation of mesenchymal stem cells into NP-like cells via 3D co-culturing with mechanical stimulation: J.Biosci.Bioeng., v. 108, p. 63–67. DOI: 10.1016/j.jbiosc.2009.02.008 16

Kim,H, J U Lee, S H Moon, H C Kim, U H Kwon, N H Seol, H J Kim, J O Park, H J Chun, I K Kwon, H M Lee, (2009)b, Zonal responsiveness of the human intervertebral disc to bone morphogenetic protein-2: Spine (Phila Pa 1976.), v. 34, p. 1834–1838. DOI: 10.1097/BRS.0b013e3181ae18ba 10

Kluba,T, T Niemeyer, C Gaissmaier, T Grunder, (2005), Human anulus fibrosis and nucleus pulposus cells of the intervertebral disc: effect of degeneration and culture system on cell phenotype: Spine (Phila Pa 1976.), v. 30, p. 2743–2748. 9

Kuisma,M, J Karppinen, M Haapea, J Niinimaki, R Ojala, M Heliovaara, R Korpelainen, K Kaikkonen, S Taimela, A Natri, O Tervonen, (2008), Are the determinants of vertebral end-plate changes and severe disc degeneration in the lumbar spine the same? A magnetic resonance imaging study in middle-aged male workers: BMC.Musculoskelet.Disord., v. 9, p. 51. DOI: 10.1186/1471-2474-9-51 6

Le Maitre,CL, P Baird, A J Freemont, J A Hoyland, (2009)a, An in vitro study investigating the survival and phenotype of mesenchymal stem cells following injection into nucleus pulposus tissue: Arthritis Res.Ther., v. 11, p. R20. DOI: 10.1186/ar2611 17

Le Maitre,CL, A J Freemont, J A Hoyland, (2004), Localization of degradative enzymes and their inhibitors in the degenerate human intervertebral disc: J.Pathol., v. 204, p. 47–54. DOI: 10.1002/path.1608 4, 5, 12

Le Maitre,CL, A J Freemont, J A Hoyland, (2005), The role of interleukin-1 in the pathogenesis of human intervertebral disc degeneration: Arthritis Res.Ther., v. 7, p. R732-R745. DOI: 10.1186/ar1732 12

Le Maitre,CL, A J Freemont, J A Hoyland, (2006), Human disc degeneration is associated with increased MMP 7 expression: Biotech.Histochem., v. 81, p. 125–131. DOI: 10.1080/10520290601005298 5

Le Maitre,CL, A J Freemont, J A Hoyland, (2007), Accelerated cellular senescence in degenerate intervertebral discs: a possible role in the pathogenesis of intervertebral disc degeneration: Arthritis Res.Ther., v. 9, p. R45. DOI: 10.1186/ar2198 4, 8

Le Maitre,CL, A J Freemont, J A Hoyland, (2009)b, Expression of cartilage-derived morphogenetic protein in human intervertebral discs and its effect on matrix synthesis in degenerate human nucleus pulposus cells: Arthritis Res.Ther., v. 11, p. R137. DOI: 10.1186/ar2808 11

Lee,CR, D Sakai, T Nakai, K Toyama, J Mochida, M Alini, S Grad, (2007), A phenotypic comparison of intervertebral disc and articular cartilage cells in the rat: Eur.Spine J., v. 16, p. 2174–2185. DOI: 10.1007/s00586-007-0475-y 15, 22

Li,X, J P Lee, G Balian, A D Greg, (2005), Modulation of chondrocytic properties of fat-derived mesenchymal cells in co-cultures with nucleus pulposus: Connect.Tissue Res., v. 46, p. 75–82. DOI: 10.1080/03008200590954104 17

Lu,ZF, B Z Doulabi, P I Wuisman, R A Bank, M N Helder, (2008), Influence of collagen type II and nucleus pulposus cells on aggregation and differentiation of adipose tissue-derived stem cells: J.Cell Mol.Med., v. 12, p. 2812–2822. DOI: 10.1111/j.1582-4934.2008.00278.x 16

Lu,ZF, D B Zandieh, P I Wuisman, R A Bank, M N Helder, (2007), Differentiation of adipose stem cells by nucleus pulposus cells: configuration effect: Biochem.Biophys.Res.Commun., v. 359, p. 991–996. DOI: 10.1016/j.bbrc.2007.06.002 15

Luo,ZJ, B B Seedhom, (2007), Light and low-frequency pulsatile hydrostatic pressure enhances extracellular matrix formation by bone marrow mesenchymal cells: an in-vitro study with special reference to cartilage repair: Proc.Inst.Mech.Eng H., v. 221, p. 499–507. DOI: 10.1243/09544119JEIM199 16

Manchikanti,L, V Singh, S Datta, S P Cohen, J A Hirsch, (2009), Comprehensive review of epidemiology, scope, and impact of spinal pain: Pain Physician, v. 12, p. E35-E70. 1

Marchand,F, A M Ahmed, (1990), Investigation of the laminate structure of lumbar disc anulus fibrosus: Spine (Phila Pa 1976.), v. 15, p. 402–410. 3

Maroudas,A, R A Stockwell, A Nachemson, J Urban, (1975), Factors involved in the nutrition of the human lumbar intervertebral disc: cellularity and diffusion of glucose in vitro: J.Anat., v. 120, p. 113–130. 3

Matsui,H, M Kanamori, H Ishihara, K Yudoh, Y Naruse, H Tsuji, (1998), Familial predisposition for lumbar degenerative disc disease. A case-control study: Spine (Phila Pa 1976.), v. 23, p. 1029–1034. 6

Matsumoto,T, M Kawakami, K Kuribayashi, T Takenaka, T Tamaki, (1999), Cyclic mechanical stretch stress increases the growth rate and collagen synthesis of nucleus pulposus cells in vitro: Spine (Phila Pa 1976.), v. 24, p. 315–319. 9

Meisel,HJ, V Siodla, T Ganey, Y Minkus, W C Hutton, O J Alasevic, (2007), Clinical experience in cell-based therapeutics: disc chondrocyte transplantation A treatment for degenerated or damaged intervertebral disc: Biomol.Eng, v. 24, p. 5–21. DOI: 10.1016/j.bioeng.2006.07.002 12

Miller,JA, C Schmatz, A B Schultz, (1988), Lumbar disc degeneration: correlation with age, sex, and spine level in 600 autopsy specimens: Spine (Phila Pa 1976.), v. 13, p. 173–178. 1

Minogue,BM, S M Richardson, L A Zeef, A J Freemont, J A Hoyland, (2010), Transcriptional profiling of bovine intervertebral disc cells: implications for identification of normal and degenerate human intervertebral disc cell phenotypes: Arthritis Res.Ther., v. 12, p. R22. DOI: 10.1186/ar2929 15, 22

Miyanishi,K, M C Trindade, D P Lindsey, G S Beaupre, D R Carter, S B Goodman, D J Schurman, R L Smith, (2006), Dose- and time-dependent effects of cyclic hydrostatic pressure on transforming growth factor-beta3-induced chondrogenesis by adult human mesenchymal stem cells in vitro: Tissue Eng, v. 12, p. 2253–2262. DOI: 10.1089/ten.2006.12.2253 16

Mwale,F, J C Iatridis, J Antoniou, (2008), Quantitative MRI as a diagnostic tool of intervertebral disc matrix composition and integrity: Eur.Spine J., v. 17 Suppl 4, p. 432–440. DOI: 10.1007/s00586-008-0744-4 6

Mwale,F, P Roughley, J Antoniou, (2004), Distinction between the extracellular matrix of the nucleus pulposus and hyaline cartilage: a requisite for tissue engineering of intervertebral disc: Eur.Cell Mater., v. 8, p. 58–63. 13, 16

Neidlinger-Wilke,C, K Wurtz, J P Urban, W Borm, M Arand, A Ignatius, H J Wilke, L E Claes, (2006), Regulation of gene expression in intervertebral disc cells by low and high hydrostatic pressure: Eur.Spine J., v. 15 Suppl 3, p. S372-S378. DOI: 10.1007/s00586-006-0112-1 10

Nockels,RP, (2005), Dynamic stabilization in the surgical management of painful lumbar spinal disorders: Spine (Phila Pa 1976.), v. 30, p. S68-S72. 2

O'Halloran,DM, A S Pandit, (2007), Tissue-engineering approach to regenerating the intervertebral disc: Tissue Eng, v. 13, p. 1927–1954. DOI: 10.1089/ten.2005.0608 9, 10

Oegema,TR, Jr., S L Johnson, D J Aguiar, J W Ogilvie, (2000), Fibronectin and its fragments increase with degeneration in the human intervertebral disc: Spine (Phila Pa 1976.), v. 25, p. 2742–2747. 4

Okuma,M, J Mochida, K Nishimura, K Sakabe, K Seiki, (2000), Reinsertion of stimulated nucleus pulposus cells retards intervertebral disc degeneration: an in vitro and in vivo experimental study: J.Orthop.Res., v. 18, p. 988–997. DOI: 10.1002/jor.1100180620 11

Osti,OL, B Vernon-Roberts, R D Fraser, (1990), 1990 Volvo Award in experimental studies. Anulus tears and intervertebral disc degeneration. An experimental study using an animal model: Spine (Phila Pa 1976.), v. 15, p. 762–767. 6

Patel,KP, J D Sandy, K Akeda, K Miyamoto, T Chujo, H S An, K Masuda, (2007), Aggrecanases and aggrecanase-generated fragments in the human intervertebral disc at early and advanced stages of disc degeneration: Spine (Phila Pa 1976.), v. 32, p. 2596–2603. DOI: 10.1097/BRS.0b013e318158cb85 5

Pazzaglia,UE, J R Salisbury, P D Byers, (1989), Development and involution of the notochord in the human spine: J.R.Soc.Med., v. 82, p. 413–415. 7

Pittenger,MF, A M Mackay, S C Beck, R K Jaiswal, R Douglas, J D Mosca, M A Moorman, D W Simonetti, S Craig, D R Marshak, (1999), Multilineage potential of adult human mesenchymal stem cells: Science, v. 284, p. 143–147. DOI: 10.1126/science.284.5411.143 13

Pockert,AJ, S M Richardson, C L Le Maitre, M Lyon, J A Deakin, D J Buttle, A J Freemont, J A Hoyland, (2009), Modified expression of the ADAMTS enzymes and tissue inhibitor of metalloproteinases 3 during human intervertebral disc degeneration: Arthritis Rheum., v. 60, p. 482–491. DOI: 10.1002/art.24291 5

Poiraudeau,S, I Monteiro, P Anract, O Blanchard, M Revel, M T Corvol, (1999), Phenotypic characteristics of rabbit intervertebral disc cells. Comparison with cartilage cells from the same animals: Spine (Phila Pa 1976.), v. 24, p. 837–844. 15

Pokharna,HK, F M Phillips, (1998), Collagen crosslinks in human lumbar intervertebral disc aging: Spine (Phila Pa 1976.), v. 23, p. 1645–1648. 4

Pope,MH, M Magnusson, D G Wilder, (1998), Kappa Delta Award. Low back pain and whole body vibration: Clin.Orthop.Relat Res., p. 241–248. DOI: 10.1097/00003086-199809000-00029 6

Puustjarvi,K, M Lammi, I Kiviranta, H J Helminen, M Tammi, (1993), Proteoglycan synthesis in canine intervertebral discs after long-distance running training: J.Orthop.Res., v. 11, p. 738–746. DOI: 10.1002/jor.1100110516 6

Rajasekaran,S, J N Babu, R Arun, B R Armstrong, A P Shetty, S Murugan, (2004), ISSLS prize winner: A study of diffusion in human lumbar discs: a serial magnetic resonance imaging study documenting the influence of the endplate on diffusion in normal and degenerate discs: Spine (Phila Pa 1976.), v. 29, p. 2654–2667. 5

Rajpurohit,R, M V Risbud, P Ducheyne, E J Vresilovic, I M Shapiro, (2002), Phenotypic characteristics of the nucleus pulposus: expression of hypoxia inducing factor-1, glucose transporter-1 and MMP-2: Cell Tissue Res., v. 308, p. 401–407. DOI: 10.1007/s00441-002-0563-6 18

Richardson,SM, J M Curran, R Chen, A Vaughan-Thomas, J A Hunt, A J Freemont, J A Hoyland, (2006)a, The differentiation of bone marrow mesenchymal stem cells into chondrocyte-like cells on poly-L-lactic acid (PLLA) scaffolds: Biomaterials, v. 27, p. 4069–4078. DOI: 10.1016/j.biomaterials.2006.03.017 15

Richardson,SM, R V Walker, S Parker, N P Rhodes, J A Hunt, A J Freemont, J A Hoyland, (2006)b, Intervertebral disc cell-mediated mesenchymal stem cell differentiation: Stem Cells, v. 24, p. 707–716. DOI: 10.1634/stemcells.2005-0205 15

Risbud,MV, T J Albert, A Guttapalli, E J Vresilovic, A S Hillibrand, A R Vaccaro, I M Shapiro, (2004), Differentiation of mesenchymal stem cells towards a nucleus pulposus-like phenotype in vitro: implications for cell-based transplantation therapy: Spine (Phila Pa 1976.), v. 29, p. 2627–2632. DOI: 10.1097/01.brs.0000146462.92171.7f 14

Risbud,MV, A Guttapalli, T T Tsai, J Y Lee, K G Danielson, A R Vaccaro, T J Albert, Z Gazit, D Gazit, I M Shapiro, (2007), Evidence for skeletal progenitor cells in the degenerate human intervertebral disc: Spine (Phila Pa 1976.), v. 32, p. 2537–2544. DOI: 10.1097/BRS.0b013e318158dea6 23

Roberts,S, B Caterson, J Menage, E H Evans, D C Jaffray, S M Eisenstein, (2000), Matrix metalloproteinases and aggrecanase: their role in disorders of the human intervertebral disc: Spine (Phila Pa 1976.), v. 25, p. 3005–3013. 5

Roberts,S, E H Evans, D Kletsas, D C Jaffray, S M Eisenstein, (2006), Senescence in human intervertebral discs: Eur.Spine J., v. 15 Suppl 3, p. S312-S316. DOI: 10.1007/s00586-006-0126-8 8

Roberts,S, J P Urban, H Evans, S M Eisenstein, (1996), Transport properties of the human cartilage endplate in relation to its composition and calcification: Spine (Phila Pa 1976.), v. 21, p. 415–420. 5

Rubin,DI, (2007), Epidemiology and risk factors for spine pain: Neurol.Clin., v. 25, p. 353–371. DOI: 10.1016/j.ncl.2007.01.004 1

Rutges,J, L B Creemers, W Dhert, S Milz, D Sakai, J Mochida, M Alini, S Grad, (2010), Variations in gene and protein expression in human nucleus pulposus in comparison with annulus fibrosus and cartilage cells: potential associations with aging and degeneration: Osteoarthritis.Cartilage, v. 18, p. 416–423. DOI: 10.1016/j.joca.2009.09.009 15, 22

Ryan,JM, F P Barry, J M Murphy, B P Mahon, (2005), Mesenchymal stem cells avoid allogeneic rejection: J.Inflamm.(Lond), v. 2, p. 8. DOI: 10.1186/1476-9255-2-8 22

Sakaguchi,Y, I Sekiya, K Yagishita, T Muneta, (2005), Comparison of human stem cells derived from various mesenchymal tissues: superiority of synovium as a cell source: Arthritis Rheum., v. 52, p. 2521–2529. DOI: 10.1002/art.21212 16

Sakai,D, J Mochida, T Iwashina, A Hiyama, H Omi, M Imai, T Nakai, K Ando, T Hotta, (2006), Regenerative effects of transplanting mesenchymal stem cells embedded in atelocollagen to the degenerated intervertebral disc: Biomaterials, v. 27, p. 335–345. DOI: 10.1016/j.biomaterials.2005.06.038 18

Sakai,D, J Mochida, T Iwashina, T Watanabe, T Nakai, K Ando, T Hotta, (2005), Differentiation of mesenchymal stem cells transplanted to a rabbit degenerative disc model: potential and limitations for stem cell therapy in disc regeneration: Spine (Phila Pa 1976.), v. 30, p. 2379–2387. DOI: 10.1097/01.brs.0000184365.28481.e3 18

Sakai,D, J Mochida, Y Yamamoto, T Nomura, M Okuma, K Nishimura, T Nakai, K Ando, T Hotta, (2003), Transplantation of mesenchymal stem cells embedded in Atelocollagen gel to the intervertebral disc: a potential therapeutic model for disc degeneration: Biomaterials, v. 24, p. 3531–3541. DOI: 10.1016/S0142-9612(03)00222-9 18

Sakai,D, T Nakai, J Mochida, M Alini, S Grad, (2009), Differential phenotype of intervertebral disc cells: microarray and immunohistochemical analysis of canine nucleus pulposus and anulus fibrosus: Spine (Phila Pa 1976.), v. 34, p. 1448–1456. DOI: 10.1097/BRS.0b013e3181a55705 15, 22

Sambrook,PN, A J MacGregor, T D Spector, (1999), Genetic influences on cervical and lumbar disc degeneration: a magnetic resonance imaging study in twins: Arthritis Rheum., v. 42, p. 366–372. DOI: 10.1002/1529-0131(199902)42:2%3C366::AID-ANR20%3E3.0.CO;2-6 6

Seguin,CA, R M Pilliar, P J Roughley, R A Kandel, (2005), Tumor necrosis factor-alpha modulates matrix production and catabolism in nucleus pulposus tissue: Spine (Phila Pa 1976.), v. 30, p. 1940–1948. DOI: 10.1097/01.brs.0000176188.40263.f9 4

Setton,LA, J Chen, (2004), Cell mechanics and mechanobiology in the intervertebral disc: Spine (Phila Pa 1976.), v. 29, p. 2710–2723. DOI: 10.1097/01.brs.0000146050.57722.2a 9

Sheikh,H, K Zakharian, R P De La Torre, C Facek, A Vasquez, G R Chaudhry, D Svinarich, M J Perez-Cruet, (2009), In vivo intervertebral disc regeneration using stem cell-derived chondroprogenitors: J.Neurosurg.Spine, v. 10, p. 265–272. DOI: 10.3171/2008.12.SPINE0835 20

Shirasawa,S, I Sekiya, Y Sakaguchi, K Yagishita, S Ichinose, T Muneta, (2006), In vitro chondrogenesis of human synovium-derived mesenchymal stem cells: optimal condition and comparison with bone marrow-derived cells: J.Cell Biochem., v. 97, p. 84–97. DOI: 10.1002/jcb.20546 16

Singh,K, K Masuda, E J Thonar, H S An, G Cs-Szabo, (2009), Age-related changes in the extracellular matrix of nucleus pulposus and anulus fibrosus of human intervertebral disc: Spine (Phila Pa 1976.), v. 34, p. 10–16. DOI: 10.1097/BRS.0b013e31818e5ddd 4

Sivan,SS, E Tsitron, E Wachtel, P J Roughley, N Sakkee, H F van der, J Degroot, S Roberts, A Maroudas, (2006), Aggrecan turnover in human intervertebral disc as determined by the racemization of aspartic acid: J.Biol.Chem., v. 281, p. 13009–13014. DOI: 10.1074/jbc.M600296200 3

Sivan,SS, E Wachtel, E Tsitron, N Sakkee, H F van der, J Degroot, S Roberts, A Maroudas, (2008), Collagen turnover in normal and degenerate human intervertebral discs as determined by the racemization of aspartic acid: J.Biol.Chem., v. 283, p. 8796–8801. DOI: 10.1074/jbc.M709885200 3

Sobajima,S, G Vadala, A Shimer, J S Kim, L G Gilbertson, J D Kang, (2008), Feasibility of a stem cell therapy for intervertebral disc degeneration: Spine J., v. 8, p. 888–896. DOI: 10.1016/j.spinee.2007.09.011 15, 18

Steck,E, H Bertram, R Abel, B Chen, A Winter, W Richter, (2005), Induction of intervertebral disc-like cells from adult mesenchymal stem cells: Stem Cells, v. 23, p. 403–411. DOI: 10.1634/stemcells.2004-0107 14

Sztrolovics,R, M Alini, P J Roughley, J S Mort, (1997), Aggrecan degradation in human intervertebral disc and articular cartilage: Biochem.J., v. 326 (Pt 1), p. 235–241. 4

Tsai,TT, A Guttapalli, E Oguz, L H Chen, A R Vaccaro, T J Albert, I M Shapiro, M V Risbud, (2007), Fibroblast growth factor-2 maintains the differentiation potential of nucleus pulposus cells in vitro: implications for cell-based transplantation therapy: Spine (Phila Pa 1976.), v. 32, p. 495–502. DOI: 10.1097/01.brs.0000257341.88880.f1 10

Urban,JP, S Smith, J C Fairbank, (2004), Nutrition of the intervertebral disc: Spine (Phila Pa 1976.), v. 29, p. 2700–2709. DOI: 10.1097/01.brs.0000146499.97948.52 4

Vadala,G, S Sobajima, J Y Lee, J Huard, V Denaro, J D Kang, L G Gilbertson, (2008)a, In vitro interaction between muscle-derived stem cells and nucleus pulposus cells: Spine J., v. 8, p. 804–809. DOI: 10.1016/j.spinee.2007.07.394 16

Vadala,G, R K Studer, G Sowa, F Spiezia, C Iucu, V Denaro, L G Gilbertson, J D Kang, (2008)b, Coculture of bone marrow mesenchymal stem cells and nucleus pulposus cells modulate gene expression profile without cell fusion: Spine (Phila Pa 1976.), v. 33, p. 870–876. DOI: 10.1097/BRS.0b013e31816b4619 15

Varlotta,GP, M D Brown, J L Kelsey, A L Golden, (1991), Familial predisposition for herniation of a lumbar disc in patients who are less than twenty-one years old: J.Bone Joint Surg.Am., v. 73, p. 124–128. 6

Vonk,LA, R J Kroeze, B Z Doulabi, R J Hoogendoorn, C Huang, M N Helder, V Everts, R A Bank, (2010), Caprine articular, meniscus and intervertebral disc cartilage: An integral analysis of collagen network and chondrocytes: Matrix Biol, v. 29, p. 209–218. DOI: 10.1016/j.matbio.2009.12.001 13

Wang,C, J D Auerbach, W R Witschey, R A Balderston, R Reddy, A Borthakur, (2007)a, Advances in Magnetic Resonance Imaging for the assessment of degenerative disc disease of the lumbar spine: Semin.Spine Surg., v. 19, p. 65–71. DOI: 10.1053/j.semss.2007.04.009 1

Wang,DL, S D Jiang, L Y Dai, (2007)b, Biologic response of the intervertebral disc to static and dynamic compression in vitro: Spine (Phila Pa 1976.), v. 32, p. 2521–2528. 9

Watanabe,T, D Sakai, Y Yamamoto, T Iwashina, K Serigano, F Tamura, J Mochida, (2010), Human nucleus pulposus cells significantly enhanced biological properties in a coculture system with direct cell-to-cell contact with autologous mesenchymal stem cells: J.Orthop.Res, v. 28, p. 623–630. 11, 22

Wei,A, H Brisby, S A Chung, A D Diwan, (2008), Bone morphogenetic protein-7 protects human intervertebral disc cells in vitro from apoptosis: Spine J., v. 8, p. 466–474. DOI: 10.1016/j.spinee.2007.04.021 10

Wei,A, S A Chung, H Tao, H Brisby, Z Lin, B Shen, D D Ma, A D Diwan, (2009)a, Differentiation of rodent bone marrow mesenchymal stem cells into intervertebral disc-like cells following coculture with rat disc tissue: Tissue Eng Part A, v. 15, p. 2581–2595. DOI: 10.1089/ten.tea.2008.0458 17

38   REFERENCES

Wei,A, H Tao, S A Chung, H Brisby, D D Ma, A D Diwan, (2009)b, The fate of transplanted xenogeneic bone marrow-derived stem cells in rat intervertebral discs: J.Orthop.Res., v. 27, p. 374–379. DOI: 10.1002/jor.20567 20

Weiler,C, A G Nerlich, J Zipperer, B E Bachmeier, N Boos, (2002), 2002 SSE Award Competition in Basic Science: expression of major matrix metalloproteinases is associated with intervertebral disc degradation and resorption: Eur.Spine J., v. 11, p. 308–320. DOI: 10.1007/s00586-002-0472-0 5

Wilke,HJ, P Neef, M Caimi, T Hoogland, L E Claes, (1999), New in vivo measurements of pressures in the intervertebral disc in daily life: Spine (Phila Pa 1976.), v. 24, p. 755–762. 9

Wuertz,K, K Godburn, C Neidlinger-Wilke, J Urban, J C Iatridis, (2008), Behavior of mesenchymal stem cells in the chemical microenvironment of the intervertebral disc: Spine (Phila Pa 1976.), v. 33, p. 1843–1849. 17

Wuertz,K, J P Urban, J Klasen, A Ignatius, H J Wilke, L Claes, C Neidlinger-Wilke, (2007), Influence of extracellular osmolarity and mechanical stimulation on gene expression of intervertebral disc cells: J.Orthop.Res., v. 25, p. 1513–1522. DOI: 10.1002/jor.20436 10

Yamamoto,Y, J Mochida, D Sakai, T Nakai, K Nishimura, H Kawada, T Hotta, (2004), Upregulation of the viability of nucleus pulposus cells by bone marrow-derived stromal cells: significance of direct cell-to-cell contact in coculture system: Spine (Phila Pa 1976.), v. 29, p. 1508–1514. 11

Yang,F, V Y Leung, K D Luk, D Chan, K M Cheung, (2009), Mesenchymal stem cells arrest intervertebral disc degeneration through chondrocytic differentiation and stimulation of endogenous cells: Mol.Ther., v. 17, p. 1959–1966. DOI: 10.1038/mt.2009.146 21

Yang,SH, C C Wu, T T Shih, Y H Sun, F H Lin, (2008), In vitro study on interaction between human nucleus pulposus cells and mesenchymal stem cells through paracrine stimulation: Spine (Phila Pa 1976.), v. 33, p. 1951–1957. 15

Yoshikawa, T, Y Ueda, K Miyazaki, M Koizumi, Y Takakura, (2010), Disc regeneration therapy using marrow mesenchymal cell transplantation: a report of two case studies: Spine (Phila Pa 1976.), v. 35, E475–480. 22

Yu,J, (2002), Elastic tissues of the intervertebral disc: Biochem.Soc.Trans., v. 30, p. 848–852. DOI: 10.1042/BST0300848 3

Zhang,Y, F M Phillips, E J Thonar, T Oegema, H S An, J A Roman-Blas, T C He, D G Anderson, (2008), Cell therapy using articular chondrocytes overexpressing BMP-7 or BMP-10 in a rabbit disc organ culture model: Spine (Phila Pa 1976.), v. 33, p. 831–838. 13

Zhang,YG, X Guo, P Xu, L L Kang, J Li, (2005), Bone mesenchymal stem cells transplanted into rabbit intervertebral discs can increase proteoglycans: Clin.Orthop.Relat Res., p. 219–226. DOI: 10.1097/01.blo.0000146534.31120.cf 18

Zhao,CQ, L S Jiang, L Y Dai, (2006), Programmed cell death in intervertebral disc degeneration: Apoptosis., v. 11, p. 2079–2088. DOI: 10.1007/s10495-006-0290-7 8

Authors' Biographies

SIBYLLE GRAD

Sibylle Grad obtained her PhD in Natural Sciences from the Federal Institute of Technology, Zurich, Switzerland. Currently she is principal investigator at the AO Research Institute Davos, Switzerland. Since 2000 she has worked in musculoskeletal research, focusing on cell based treatment approaches for musculoskeletal regeneration. Main research activities include tissue engineering of articular cartilage and intervertebral disc, in vitro differentiation of mesenchymal stem cells, bioreactor culture systems, and characterization of intervertebral disc cell phenotypes. She has published and presented widely and is actively involved in different collaborative research projects worldwide.

DAISUKE SAKAI

Daisuke Sakai is currently an assistant professor at Department of Orthopaedic Surgery, Surgical Science and Center for Regenerative Medicine, Tokai University School of Medicine, Kanagawa, Japan. Dr. Sakai's current research focuses on maintenance of homeostasis and intrinsic stem cell system of the intervertebral disc cells and stem cell therapy, TGFb/SMADs signaling, low-intensity pulsed ultrasound applications and clinical spine surgery. Dr. Sakai has given over 110 presentations, invited lectures and seminars, in universities, professional societies, international conferences, and government organizations around the world. He actively participated in grant review for NIH and other funding agencies, and reviews of papers for many journals.

JOJI MOCHIDA

Joji Mochida presently serves as Professor and Chairman at Department of Orthopaedic Surgery, Tokai University School of Medicine, Kanagawa, Japan. Dr. Mochida's main research interest is in biological treatments to retard the progressive nature of intervertebral disc degeneration. After a decade of laboratory and clinical research, he has conducted a clinical trial which transplants activated nucleus pulposus cells into mildly degenerated discs. Dr. Mochida is also a surgeon with experience in various spinal surgeries. Dr. Mochida also serves as an associate Dean of the Medical School and breaks time to educate many students and young doctors and researchers

MAURO ALINI

Mauro Alini graduated in Chemistry from the University of Lausanne (Switzerland) in 1983. Since then he has been involved in connective tissue research, starting form his Ph.D. research work, which focused on the isolation and characterization of proteoglycans extracted from both normal human mammary gland and carcinomas thereof. In September 1988, he joined the Joint Diseases Laboratory (under Dr. A. R. Poole's direction) at the Shriners Hospital in Montreal to work on growth plate tissue before and at the time of cartilage matrix calcification during endochondral bone formation. In January 1995, he was appointed as an Assistant Professor at the Division of Surgery of the McGill University, and head of the Biochemistry Unit of the Orthopaedic Research Laboratory, working to develop new biological approaches to treating intervertebral disc damage. Since July 2000, he is in charge of the Musculoskeletal Regeneration Program at the AO Research Institute (Davos, Switzerland), focusing on cartilage, bone and intervertebral disc tissue engineering.

CHAPTER 2

Recent Advances in Biomaterial Based Tissue Engineering for Intervertebral Disc Regeneration

Sunil Mahor, Estelle Collin, Biraja Dash, and Abhay Pandit
National University of Ireland, Galway, Ireland

David Eglin
AO Research Institute Davos, Switzerland

2.1 INTRODUCTION

Low back pain (LBP) is a predominant cause of disability, with significant socio-economic consequences (Anderson, J., 1986; Borenstein, D., 1992; Ehrlich, G., 2003). LBP is particularly prevalent in 20- to 50-year-olds, although it is more severe in older patients (Biering, S., 1982). LBP has emerged as the most expensive healthcare problem in the United States, with reported annual costs, direct and indirect, ranging from \$20 to \$100 billion as reported in 2003 (Aminian et al., 2003). LBP is generally associated with degeneration of the intervertebral disc (IVD) (Kelsey and White, 1980; Deyo and Tsui-Wu, 1987; Yasuma et al., 1990; Kuslich et al., 1991; Frymoyer and Cats-Baril, 1991; Schwarzer et al., 1995; Waddell, G., 1996; O'Neill et al., 2002). However, this causal relationship is under constant consideration because no absolute link has been found between clinical symptoms and degenerative disc disease (DDD) (Boos et al., 2000; Elfering et al., 2002; An et al., 2003a; Snook, S., 2004). Changes in the matrix composition and deterioration of biomechanical properties, abnormal mechanical loading, genetic predisposition, reduced cell activity, or any combination of the above may induce DDD (Lyons et al., 1981; Miller et al., 1988; Lipson and Muir, 1981; Guiot and Fessler, 2000; Hutton et al., 1998; Natarajan et al., 1994; Lotz and Chin, 2000).

With the design and development of biomaterials for therapeutic interventions, a wide range of clinical disorders is being targeted. Recent studies of cell-matrix interactions, cell–cell signaling, and organization of matrix components are increasing the demands for biomaterials and continue to create interest in developing new biomaterials and improving the performance of existing medical-grade biopolymers. This chapter describes biologically derived and synthetic biomaterials that have been previously used or have been under consideration for use in intervertebral disc (IVD) regeneration applications. Particular emphasis is put on advances in nucleus pulposus (NP) regenerative strategies since it is believed that in most cases degenerative changes originate in this central part of the disc.

2.2 INTERVERTEBRAL DISC: STRUCTURE, FUNCTION AND CELLS

2.2.1 ANATOMY OF THE INTERVERTEBRAL DISC

The IVDs are the soft tissues located between each of the 24 cervical, thoracic, and lumbar vertebrae of the spine. They lie between the vertebral bodies and are separated from them by a hyaline cartilage endplate. The IVD is an avascular structure. Poor blood supply and low cellularity may contribute to nutritional and degenerative problems in the intervertebral disc. The intervertebral discs vary in size and shape with spinal level, but the general structure and composition of the intervertebral discs are constant along the length of the spine (Sebastine and Williams, 2007). The structure of each intervertebral disc unit has three main components: the *nucleus pulposus* (NP) – the gelatinous, hydrated center of the intervertebral disc; the *annulus fibrosus* (AF) – a fibrous ring which surrounds the *nucleus pulposus*; and the end plates, which are situated above and below each intervertebral disc, adjacent to the vertebrae (Figs. 2.1 and 2.2) (Sun et al., 2001; Humzah and Soames, 1988). End plates separate the vertebral bodies to the IVDs and are responsible for nutrition by transportation of biomolecules in and out of the disc.

The AF consists of a series of loosely connected concentric layers (lamellae) of highly oriented collagen (mainly type I) fibers that enclose the NP (Schollmeier et al., 2000; Roberts et al., 1989; Ebara et al., 1996). Within these lamellae, the collagen fibers lie parallel to each other at an angle of approximately 60° to the spine axis and are oriented in opposite directions in successive layers (Walmsley, R., 1953; Inerot and Axelsson, 1991; Hickey, D., 1980). The peripheral area of the AF attaches to the posterior longitudinal ligament and inserts into the vertebral body via Sharpey's fibers. The innermost region merges horizontally with the NP, and the collagen fibers continue vertically through the endplate. The cells of the annulus fibrosus vary from inner to outer layer. The inner annulus fibrosus is populated with cells similar to nucleus pulposus cells. These cells are round and chondrocyte-like. Outer annulus fibrosus cells are more fibrocyte-like and elongated, with their major axes aligned along the collagen fiber direction (Setton and Chen, 2004). These cells produce predominantly type I and type II collagen. The cell density in the annulus fibrosus is 9×10^6 cells/cm^3, which is very low compared to other tissues.

The NP comprises a more random network of collagen fibrils (predominantly type II) that are loosely embedded in a proteoglycan (PG)-rich gelatinous matrix. This tissue is confined within the annular lamellae and exhibits a high affinity for water due to its highly negative sulfated charge.

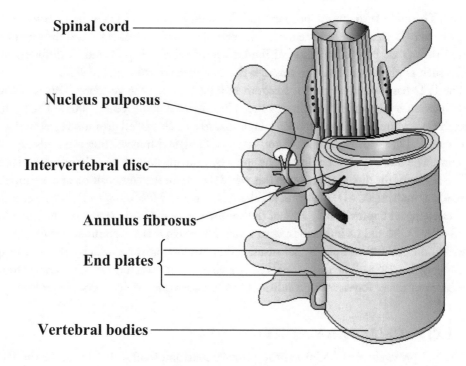

Spinal cord

Nucleus pulposus

Intervertebral disc

Annulus fibrosus

End plates

Vertebral bodies

Figure 2.1: Anatomy of the intervertebral disc.

The overall structure of the disc is analogous to that of a tire, in which air pressure (hydrated NP) keeps the rubber tire (AF) inflated. The cell density in the nucleus pulposus is very low compared to other tissues. Whereas other relatively acellular tissues such as cartilage have cell densities of 14×10^6 cells/cm^3, nucleus pulposus cell density is nearly an order of magnitude lower at 4×10^6 cells/cm^3 (Oegema, T., 1993). In very early life, the cells in the nucleus pulposus are derived from fetal notochordal cells. These cells disappear completely by early adulthood and are replaced by a lesser number of round cells that resemble the chondrocytes found in tissues subjected to compressive loading, such as articular cartilage (Coventry et al., 1945). These cells produce type II collagen and aggrecan predominantly.

End plates lie at the cranial and caudal interface between the IVD and the vertebral body. These plates are connected directly to the lamellae in the inner region of the AF, but there is no sign of collagenous connections with the underlying vertebral body (Hukins, D., 1988). They consist of translucent, hyaline cartilage, and an osseous component (Moore, R., 2000). The central region of the end plates is thinner and more permeable than the thicker, impermeable peripheral region. These structures contain nutrition pathways that connect to vascular buds lying in the medullary space (Nachemson et al., 1970; Oki et al., 1996) and facilitate diffusion of molecules into and out of the disc. The end plates are the primary path for fluid flow to and from the intervertebral disc. Capillary density is greatest in the center of the end plate and decreases toward the outer annulus fibrosus (Urban et al., 2004). Nutrition of the intervertebral disc is provided by diffusion of small molecules such as glucose and oxygen through the end plates (Urban et al., 1982).

The IVD has a multi-degree of freedom of joint between the vertebral bodies, whereby its primary function is to confer flexibility to the spine (Nordin and Weiner, 2001). The interaction between the intervertebral disc components is similar to a thick-walled pressure vessel, and it allows the intervertebral discs to act as shock absorbers, absorbing and transmitting the loads experienced by the spine. The disc imparts stability to the spine by resisting compressive, bending, and torsional loads. The ability of the disc to function in this way depends on the composition and organization of its extracellular matrix (ECM) (Bayliss et al., 1988; Urban and McMullin, 1985; Urban et al., 2000). The NP can support approximately 70% of the compressive axial load exerted on the spine because of its PG-rich ECM and its interaction with water. However, it is the confinement of this gel-like tissue within the AF and end plates that sustains this swelling pressure, which balances the applied loads (Broberg, K., 1983). The disc is inherently a viscoelastic structure, which allows it to behave as a flexible joint under low-loading conditions but becomes quite rigid under higher loads.

2.2.2   EXTRACELLULAR MATRIX

Collagens and proteoglycans (PGs) are the primary macromolecules that make up the ECM of the IVD (Urban and Maroudas, 1980; Bushell et al., 1977). The matrix composition of the disc varies considerably between the AF and the NP. Progressing from the AF to the NP, there is a significant increase in PG content, with a concomitant drop in total collagen content. In healthy discs, collagens account for approximately 70% of the dry weight of the AF and 20% of the dry weight of the NP. The PG content changes from only a small percentage of the dry weight of the AF to approximately 50% of the dry weight of the NP (Buckwalter, J., 1995; Eyre, D., 1979, 1988). The collagen content of the NP consists primarily of collagen type II (approximately 80%) and also small amounts of type VI, IX, and XI. The AF, alternatively, consists mainly of collagen type I, with smaller quantities of type V, VI, IX, and XI. Small quantities of collagen type III are also located in the NP and inner AF (Beard et al., 1981). These minor collagens are thought to influence the interaction between collagen fibrils and PGs and to regulate collagen type I and II fibrillogenesis in the disc (Roberts, S., 1991; Neame et al., 2000).

The largest and most important PG in the disc matrix is aggrecan, which is made of a protein core with glycosaminoglycans (negatively charged chondroitin sulfate and keratin sulfate) attached (Ghosh et al., 1977; Kitahara, H., 1979). The chondroitin sulfate molecule plays a crucial role in imbibing water, which, in turn, gives the disc its resilient compressive strength (Ohshima et al., 1995). In the central region of the NP, the aggrecan molecules accumulate on a hyaluronan strand to form structures called aggregates (Buckwalter et al., 1985, 1989). Small PGs (decorin, lumican, biglycan, and fibromodulin) that are also present in the disc are thought to influence cellular deposition and collagen assembly (Sztrolovics et al., 1999; Hayes et al., 2001).

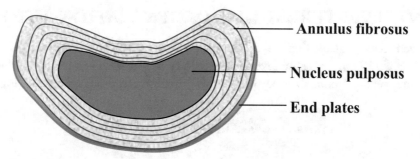

Figure 2.2: Illustration of the various components of the intervertebral disc (IVD). Each IVD is located between vertebral bodies of the spinal column. End plates are the top and bottom layers of tissues which separate the vertebral bodies to the IVDs and are responsible for nutrition by transportation of molecules in and out of the disc. Annulus fibrosus is the external lamellar tissue layers which surround the central nucleus pulposus (gel-like) tissue.

Elastin is also found throughout the IVD (Hukins, D., 1988; Yu, J., 2002). These elastic fibers are randomly oriented in the NP and restore the shape of the tissue after deformation. Elastin is more densely located in the interlaminar space of the AF where it lies parallel to the collagen fibers and forms cross-bridges between the lamellae in the outer AF region.

2.2.3   CELLS OF THE INTERVERTEBRAL DISC

The IVD has a lower cell density than other tissues, with the total cell population occupying less than 1% of the total disc volume (Urban et al., 2000; Bibby et al., 2001). The outer AF contains fibroblast-like cells that are elongated and position themselves parallel to the predominant collagen fibril orientation (Baer et al., 2003; Bruehlmann et al., 2002; Errington et al., 1998). The inner AF and cartilaginous end plates contain chondrocyte-like cells. Up to approximately 10 years of age, the NP consists of notochordal and chondrocyte-like cells (Walmsley, R., 1953; Wolfe et al., 1965; Maldonado and Oegema, 1992). Eventually, the notochordal cell population becomes depleted, and the NP is reported to retain chondrocyte-like cells only. The notochordal cells, which are derived from the notochord, can be distinguished morphologically from chondrocyte-like cells by their larger size (Maldonado and Oegema, 1992), large intracellular glycogen deposits, and poorly developed mitochondria, which are surrounded by rough endoplasmic reticulum (Trout et al., 1982a,b). Larger clusters of notochordal cells are found embedded in the ECM of young, healthy NPs. While co-existing with chondrocyte-like cells in the NP, they are considered to play a regulatory role in the biosynthetic activity of chondrocyte-like cells via the secretion of soluble signal factors and possible physical cell-to-cell contacts. The main function of the cells is to preserve the metabolic homeostasis of the disc by balancing the synthesis of ECM macromolecules and proteases (Urban et al., 2000; Bibby et al., 2001; Gruber and Hanley, 2003a,b).

2.3   INTERVERTEBRAL DISC DEGENERATION: ETIOLOGY

It is believed that degeneration of IVDs commences as early as the second decade of life (Nerlich et al., 1997). There are several contributory factors such as mechanical, biochemical, nutritional, and genetic factors that may induce this progressive, age-related disease (Fig. 2.3).

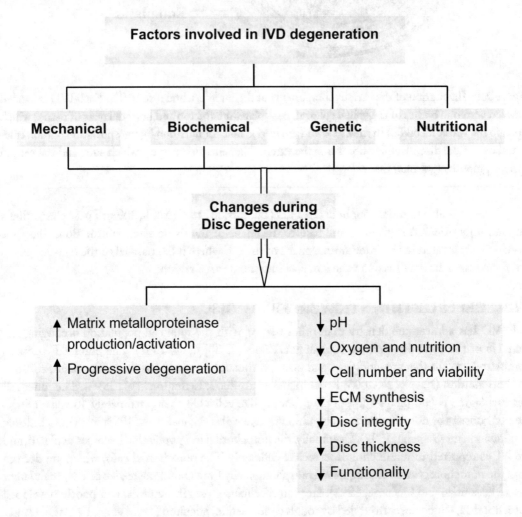

Figure 2.3: Schematic representation of various underlying factors and subsequent changes during IVD degeneration.

Mechanical stresses regulate the remodeling of connective tissues (Wolff's Law), but abnormal loading of the discs may result in fatigue failure of the ECM that leaves the cells of the disc more vulnerable to mechanical loads (Adams et al., 2000). In addition, specific genes may be triggered that lead to increased production and activation of matrix metalloproteinases (MMPs) (Rannou et al., 2001).

Calcification of the cartilaginous end plates impairs diffusion of glucose, oxygen, and amino acids into the disc (Moore, R., 2000; Horner and Urban, 2001; Turgut et al., 2003; Baogan et al., 2001; Boos et al., 2002) and thus creates an acidic and low-oxygenated extracellular environment un-suited for maintaining cell viability and macromolecular synthesis (Cassinelli et al., 2001; Bibby et al., 2001; Ohshima and Urban, 1992; Buckwalter, J., 1995; Norcross et al., 2003; Antoniou et al., 1996). An imbalance between matrix-degrading enzymes and their inhibitors may lead to yet further de-generation (Goupille et al., 1998; Crean et al., 1997; Kanemoto et al., 1996; Grange et al., 2001).

Some individuals may be predisposed to developing DDD because of their genetic make-up (MacGregor et al., 2004). There is a link between DDD and certain polymorphisms and muta-tions in genes that are associated with the production of ECM proteins. Mutations known to affect collagen type II genes in humans can lead to end-plate deformities (Rannou et al., 2003). Polymor-phisms in the aggrecan (Kawaguchi et al., 1999) and vitamin D receptor gene (Videman et al., 1998) are linked with an increase in the risk of an individual developing DDD. In summary, loss of disc mechanical integrity with nucleus pulposus dehydration and fibrosis results in a cascade of loss of normal disc mechanical function, tissue injury, and injury response. The time course of the injury and its response determines the clinical manifestation of degenerative disc disease.

2.4   CURRENT TREATMENTS FOR DEGENERATIVE DISC DISEASE

Current treatment modalities involve conservative management (medication and physical therapy) or surgical intervention (spine fusion, total disc replacement, or nucleus pulposus (NP) replace-ment) (An et al., 2003a; Kandel et al., 2008).

Fusion of the vertebral bodies above and below the implicated disc are usually opted for when gross instability of the spine segment is detected (Bao et al., 1996). Operative fusion (arthrodesis) involves placement of a bone graft within the disc space post-discectomy, with further stabilization of the joint being achieved using spinal instrumentation (Langrana et al., 1994). A study from The Healthcare Cost and Utilization Project reported that 297,883 spinal fusions were performed in the United States in 2003 (Healthcare Cost and Utilization Project, 2003). Interbody fusion has resulted in successful clinical outcomes with success in alleviating pain by removing the disc and eliminating any motion through the joint (Burkus, J., 2002; Kwon et al., 2003; Moore et al., 2002; Burkus et al., 2004; Blumenthal et al., 2002). However, long-term follow-ups on patients that have received fusion surgeries indicate degeneration and altered biomechanics at adjacent spine segment levels.

In an attempt to preserve mobility of the joint and reduce the effect on adjacent levels, other re-placement devices such as total disc replacement (TDRs) and nucleus pulposus replacement (NPRs) have been developed (Sagi et al., 2003; Darwis et al., 2002; Huang and Sandhu, 2004). The most widely implanted TDR device is the SB Charite III (DePuy, Johnson & Johnson, Raynham, MA).

This device has produced satisfactory clinical outcomes (Blumenthal et al., 2005; McAfee et al., 2005; Lemaire, J., 1997; Zeegers et al., 1999) and received Food and Drug Administration (FDA) approval for treatment of single-level disease at the L4-5 or L5-S1 levels in 2004. Other TDR devices that are currently undergoing FDA Investigational Device Exemption (FDA IDE) trials in the United States are ProDisc™ (Spine Solutions/Synthes, New York, NY), Flexicore™ (Spinecore/Stryker Spine, Allendale, NJ), and Maverick™ (Medtronic Sofamor Danek, Minneapolis, MN) (Bertagnoli et al., 2005; Zigler, J., 2003).

As an alternative to TDRs, research has focused on the development of nucleus pulposus replacements (NPRs), which allow redistribution of the loads to the remaining native structures of the disc (i.e., AF and end plates), as well as providing a less-invasive surgical procedure. The Prosthetic Disc Nucleus device (PDN®, Raymedica, Bloomington, MN) (Eysel et al., 1999; Wilke et al., 2001; Klara and Ray, 2002) has achieved the certificate of Conformité Europe and is used commercially in more than 30 countries, with FDA IDE trials ongoing in the United States. Other NPR devices that are at different stages of investigation (Di Martino et al., 2005a) include Newcleus® (Zimmer Spine, Minneapolis, MN), Aquarelle® (Stryker Spine, Allendale, NJ), Neudisc® (Replication Medical Inc., New Brunswick, NJ), Regain® (EBI, Parsippany, NJ), and Intervertebral Prosthetic Disc (IPD®, Dynamic Spine, Mahtomedi, MN). Although with NPR devices, the initial surgical procedure (and any revision) is less invasive than fusion and TDR surgery, problems have arisen with PDN implant extrusion and end-plate deformities (i.e., subsidence) with the PDN® and Newcleus® devices (Shim et al., 2003).

While these surgical methods allow symptomatic relief by removal of disc tissue, they do not specifically address the underlying biological problem (Boyd and Carter, 2006). This need has prompted the adoption of an integrative mode of research that focuses on a more biological and regenerative approach to treating degenerated discs. DDD is believed to originate in the NP region of the disc (Adler et al., 1983) because this is the region of tissue that shows the most dramatic transformation with age. Thus, most tissue engineering studies are directed toward treatment of the NP. Employment of a tissue engineering strategy is intended to manipulate the disc biology and inhibit further degeneration by introducing therapeutic agents (cells, scaffolds, and growth factors) into the NP to restore disc functionality.

2.5 BIOMATERIALS FOR ENGINEERING OF INTERVERTEBRAL DISC TISSUE

IVD tissue engineering presents the opportunity to restore the functionality by repairing or replacing the degenerated tissue. In repairing the IVD, the aim is to induce regeneration of the tissue *in situ* via biological manipulation, whereas replacing the IVD requires development of a functional tissue unit *in vitro* and its implantation in the body. The physiological properties of the disc are linked to the composition of its ECM. Thus, tissue engineering methods tend to focus on techniques directed toward replenishing the ECM components of the disc in order to restore disc function.

IVD regeneration approaches are broadly classified based on the three principal components (cells, biomaterials and signals), which may be used independently or incorporated in combinatorial form (Fig. 2.4).

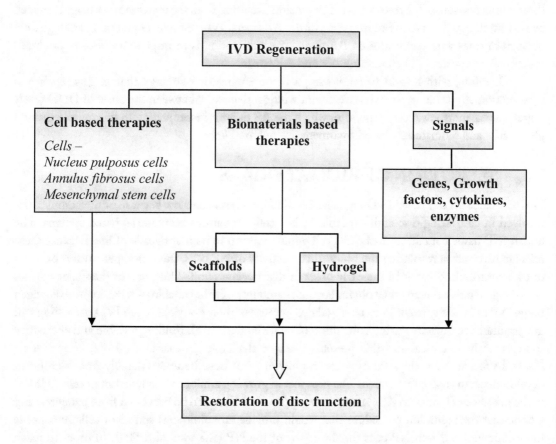

Figure 2.4: Schematic representation of various therapies for IVD regeneration.

From a clinical perspective, the degeneration state of the disc (the viability and metabolic state of the resident disc cells and the structure of the surrounding ECM) dictates what treatment strategy (cell implantation, injection of signals, or implantation of an *in vitro* developed IVD tissue) is optimal (An et al., 2003b). In the early stages of degeneration when the resident cells of the NP cease to repair the ECM and the tissue undergoes fibrotic changes, it is hypothesized that the use of growth factor (GF) therapy alone may suffice in stimulating the cells to synthesize normal ECM. Several issues with the growth factor delivery are half life, release pattern of the biomolecules, as well as dependence on single bolus delivery to succeed. Controlled release growth factor based strategies are likely to lead to therapeutically relevant results.

If DDD has reached an advanced stage, the stability of the spine segment is extensively reduced, and biological tissue engineering treatments are no longer an option. Targeting the NP for IVD tissue engineering practices is dependent on having a functional AF tissue so that it can sustain the swelling pressure of the restored NP. The tension (resulting from the increased disc height) created by this swelling will serve to induce repair of the AF fibers. Where tissue engineering strategies are pursued in cases with compromised AF tissue, the only option is to implant an *in vitro* produced functional composite IVD containing NP and AF tissue.

Therefore, with respect to employing a tissue engineering strategy that targets repair and regeneration of NP tissue, there is a treatment window that exists between the onset of DDD (early stage) and the more advanced stage, after which the AF tissue is incompetent. In this middle-ground phase, cell- and biomaterials-based treatment options are likely.

2.6   SELECTION OF BIOMATERIALS

Progress made on cell based IVD regeneration therapies has been slow because of the complexities involved in cultivating disc and, particularly, NP cells on various biomaterial based systems. The human NP tissue contains notochordal cell populations up to its first decade of life. Because these cells are involved in regulating the biosynthetic activity of NP cells, their disappearance is believed to be a contributing factor in the early onset of disc degeneration. Thus, one of the major hurdles of NP regenerative therapy is to obtain adequate amounts of cells capable of repairing the damaged tissue. NP cell activity needs to be up-regulated because of their low yields from IVD harvesting and low proliferative capacity during cultivation (Gruber et al., 1997a). Finding a viable and alternative source of cells is critical and this presents a major challenge (Aguiar et al., 1999; Hunter et al., 2003). In recent years, the field of tissue engineering has benefited considerably from significant developments in stem cell research. The prospect of growing adult mesenchymal stem cells (MSCs) in the presence of biomaterials is an attractive approach. MSCs can be harvested from patients using a bone marrow aspiration procedure that would provide an autologous source of cells susceptible to be differentiated into NP cells for the repair of the NP (Brisby et al., 2004). In order to reach this goal, the biomaterials and their selection play an important role. Different cell types including NP cells, mesenchymal stem cells, and other chondrocytic cells have been investigated in cell based tissue engineering approaches and are summarized in Table 2.1.

Biodegradable polymers are currently the vast majority of the materials used in IVD and NP repair (Table 2.2). Important criteria for the selection of biodegradable polymers for tissue repair include (i) manufacturing feasibility; (ii) the process from which a polymer can easily be converted into the final product design; (iii) mechanical properties that adequately address short-term function and do not interfere with long-term function of the biological tissue; (iv) low or negligible toxicity of degradation products, in terms of both local tissue response and systemic response; and (v) drug delivery compatibility in applications that call for release or attachment of active compounds (Pachence et al., 2007). In tissue engineering or regenerative therapy, a matrix or scaffold is most often desirable to permit cell adhesion and proliferation for the necessary multiplication

Table 2.1: Various cell-based tissue engineering approaches for IVD regeneration.

Biomaterials used	References
I. NP cells based tissue engineering	
Alginate	Aguiar et al., 1999; Thonar et al., 2002; Gaetani et al., 2008
Collagen type I	Bron et al., 2009
Hyaluronan (HA)	Haberstroh et al., 2009
Fibrin	Bertram et al., 2005
Chitosan	Mwale et al., 2005
Carboxymethylcellulose	Reza and Nicoll, 2010
Fibrin	Stern et al., 2000
Poly(vinyl alcohol) (PVA), Poly(vinyl pyrrolidone) (PVP)	Thomas et al., 2003
N-doped plasma-polymerized ethylene	Mwale et al., 2008
Chondroitin-6-sulate	Yang et al., 2005
Collagen type II	Halloran et al., 2008
Poly(L-lactide-co-glycolide)	Hong et al., 2009;
Gelatin	Yang et al., 2005
II. Mesenchymal stem cells based tissue engineering	
Atelocollagen	Sakai et al., 2006
Hyaluronan	Sato et al., 2003
KLD-12 peptide	Crevensten et al., 2004
Chitosan-glycerophosphate	Sun and Zheng, 2009
III. Chondrocyte based tissue engineering	Richardson et al., 2008
Poly(L-lactide-co-glycolide)	Hong et al., 2009; Jung et al., 2008
Amidic derivative of alginate	Leone et al., 2008

of cells and to retain extracellular matrix for its accumulation and the synthesis of new functional tissue (Wakatsuki and Elson, 2003; Gruber and Hanley, 2003b). The scaffold is then referred to as a functional template that guides the cellular remodeling process and can potentially provide the cells with temporary protection from unfavorable local implantation environments (Gruber et al., 2004). The scaffold can also function as a delivery vehicle for drugs and allows the implantation of cells in the intended site in addition to increasing their level of local retention thereafter.

Thus, when targeting an IVD and specifically NP regeneration therapy, the selection of a suitable biodegradable polymer may obviously be related to the therapeutic approach considered and the specific properties and functions of the damaged tissue. The highly hydrated nature of the NP tissue is similar to that of hydrogels. Hydrogels are 3D networks of hydrophilic polymers held together by association bonds such as covalent bonds and weaker cohesive forces such as hydrogen and ionic bonds and intermolecular hydrophobic association. These networks are able to retain a large quantity of water within their structure without dissolving and are the most studied and prime candidates to serve as cell carrier for NP regeneration (Nguyen and West, 2002). Hydrogels and other polymeric scaffolds are classified in several ways, but the simplest classification is in materials obtained from a natural product or biopolymer and materials made of a synthetic polymer.

2.6.1   BIOPOLYMERS

Biopolymers used for *in vitro* and *in vivo* cell based NP therapy studies are numerous and range from the most widespread cell carriers, collagen type I, calcium alginate, hyaluronan, fibrin, chitosan, to the less common KLD-12 peptide, carboxymethyl cellulose, collagen type II. Most of the biopolymers are employed in a highly hydrated form or hydrogel, in which cells can be encapsulated prior or during gelling of the polymeric structure. Most of the biopolymer hydrogels are susceptible to enzymatic degradation in the presence of cells and thus provide a bioresorbable temporary three-dimensional (3D) matrix. Biopolymer based scaffolds with pre-formed macro- and/or micro-porous structure have also been used (Yang et al., 2005). Hundred micron pores were created in gelatin/chondroitin-6-sulfate scaffolds to allow seeding of human NP cells within the bulk of the cross-link structure. Cell survival up to 12 weeks was reported, but no comparison with other matrices was made. The combination of biopolymers or biopolymer and synthetic polymers within a composite scaffold is also a frequent approach to recapitulate the mechanical and structural properties of the NP or the whole IVD (Sha'ban et al., 2008; Mauth et al., 2009; Mizuno et al., 2004).

2.6.2   SYNTHETIC POLYMERS

Synthetic biodegradable polymers have been used more scarcely than biopolymers in NP cell therapy. Their properties are usually well controlled and variations between batches or immune responses are less of a concern. Synthetic polymers (e.g., polyvinyl alcohol (PVA) hydrogel) with mechanical properties matching those of the NP are of strong interest for the tissue replacement, but they have seldom been combined with cells in a regenerative approach (Wang and Campbell, 2009; Allen et al., 2004; Thomas et al., 2003). Cells combined with biodegradable poly(glycolic-co-lactic

Table 2.2: Biodegradable polymer scaffold based tissue engineering for IVD regeneration *Continues.*

Scaffolds	Cell type and source	Methods	Outcomes	References
Alginate beads	NP and AF cells from adolescent rabbit IVDs	• Cells were seeded within alginate beads at 2×10^6 cells/ml. • Constructs were cultured *in vitro* for 14 days in DMEM/F-12 medium supplemented with 10% fetal bovine serum (FBS), ascorbate and gentamicin.	• Both cell types maintained their characteristic phenotype within the scaffold. • Histological analysis revealed that AF cells, which remained as single cells throughout the scaffold produced significantly more PG-rich pericellular matrix than NP cells which had arranged themselves in cell clusters.	Chiba *et al.*, 1997
Alginate beads	Porcine IVDs cell (NP possesses a large majority of notochordal cells	• Cells were seeded at 1×10^6 cells/ml in alginate beads. • Constructs were cultured *in vitro* for 16 weeks in Hams F-12 medium containing FBS, Pen/Strep and fungizone.	• Cells maintained their phenotype. • Elevated gene expression levels of collagen type I and II were observed with increased culture time. • No mechanically functional matrix was observed (compression and shear testing).	Baer *et al.*, 2001
Collagen type II (Sigma, St. Louis, MO) pre-coated Millipore CM® filter inserts	NP cells from nine month old sheep lumbar IVDs	• Cells were seeded within the scaffolds at 3.3×10^5 cells/cm². • Cell seeded constructs were cultured *in vitro* in DMEM supplemented with 10% FBS, which after five days was increased to 20%. On day seven, ascorbic acid was also added to the medium.	• DNA content had decreased after four weeks. • Cell density was comparable to native NP tissue. • NP cells maintained their phenotype, forming layer of tissue comprising PGs and type II collagen with comparable concentrations to native NP tissue. • By week 12 there were signs of deterioration.	Sun *et al.*, 2001
Freeze dried gelatin sections of Gelfoam® (Upjohn Co., Kalamazoo, MI)	IVD cells from Sand rat *Psammomys obesus*	• Autologous disc cells were seeded within the scaffold at 10,000–21,000 cells/scaffold. Cell seeded constructs were implanted into the sand rat disc degeneration model.	• Cells maintained their normal morphologies, exhibiting spindle-shape and rounded morphology in the AF and NP, respectively. • Cells were observed to fully integrate into the IVD and after eight months were surrounded by normal ECM.	Gruber *et al.*, 2002
Composite scaffold composed of Collagen type I and hyaluronan in a 9:1 ratio respectively	NP and AF cells from two-three year old bovine steers	• Cells were seeded at 2×10^6 cells/ml in composite scaffolds. • The cell seeded constructs were cultured *in vitro* in DMEM supplemented with 10% FBS for 60 days on an orbital shaker. • The effects of growth factors were also examined and will be discussed in the next section.	• No significant difference between the capacity for NP and AF cells to synthesise ECM macromolecules; aggrecan, small leucine rich proteoglycans (SLRP) and fibrillar collagens. • Main limitations of the scaffold design concerned the failure to accumulate the PG produced by the cells and any PG that was retained was not uniformly distributed throughout the scaffold.	Alini *et al.*, 2003

Table 2.2: *Continued.* Biodegradable polymer scaffold based tissue engineering for IVD regeneration *Continues.*

Scaffold material	Cell source	Method	Results	Reference
Atelocollagen type II (0.3%) (KOKENCELLGEN; Koken Co., Ltd., Tokyo, Japan)	Mesenchyme (MSCs) stem cells of Rabbits	• MSCs that were infected with Ad-lacZ expressing E. coli lacZ gene, were seeded within the scaffolds at 1×10^6 cells/ml. • Autologous cell-seeded scaffolds in solution form were injected into rabbit disc degeneration models (NP tissue aspiration).	• The cells survived post-implantation and differentiated into spindle-shaped cells that arranged themselves in longitudinal layers and synthesized significant amounts of PG. • Histology (eight weeks) revealed that groups with MSC seeded scaffolds had preserved annular structure and decelerated degeneration compared to control groups (no treatment).	Sakai et al., 2003
Poly-D, L-lactide (PDLA) beads. Demineralized bone matrix (DBM) (bovine). Gelatin microcarriers	Notorchodal and NP cells from 6-8 month old porcine IVDs	• Cells were seeded in scaffolds at 5×10^6 cells/ml scaffold. • Cell seeded scaffolds were cultured for up to four weeks in a Cel-Gro® Tissue Culture Rotator.	• Cells failed to adhere to PDLA beads. Cells on BDM and gelatin scaffolds assumed fibroblast-like morphology. • Cells on the BDM surface showed increased and decreased gene expression levels for type I and II collagen respectively compared to the gelatin scaffold.	Brown et al., 2004
Small intestine submucosa (SIS) decellularised scaffold	AF and NP cells of human degenerated IVDs	• Cells were seeded onto SIS scaffold material. • Cell seeded constructs were cultured *in vitro* for up to three months.	• During the initial stage of culturing, more than 70% of cells had attached to the scaffold. • DNA content decreased (three months), but the cells remained metabolically active-expression levels of collagen type I, II, X, transcription factor Sox-9 and aggrecan mRNA.	Le Visage et al., 2004
15% Hyaluronan scaffold	MSCs of adult rat bone marrow	• Cells were embedded in the scaffolds at 1×10^7 cells/ml. • Allogenic cell seeded scaffolds were injected into healthy rat disc models.	• Significant decrease in cell number at day seven (may be due to initial cytotoxic level of hyaluronan, or expulsion of cells from the disc). • Injection site in the AF was closed after 28 days, and tissue resembling native NP tissue had formed. • Increase in number of labeled MSCs at day 28. • Increased disc height observed with discs that were injected with cell-seeded scaffolds.	Crevensten et al., 2004
Two types of collagen sponges (Gelfoam Pharmacia and Upjohn Co., USA) and (CollaPlug_, Calcitek, USA). Collagen gels (Cell Prime, Cohesion Technologies, Inc., USA), agarose, alginate and fibrin gels	AF cells of human IVDs with Thompson grade II, III and IV	• Scaffolds seeded with AF cells were cultured for ten days *in vitro* in modified Minimal Essential Media (MEM) with Earle's salts, supplemented with L-glutamine, non-essential amino acids, Pen/Strep and 20% FBS.	• Both forms of collagen sponge resulted in the most profuse levels of cell adhesion, ECM production and expression of genes for type I and II collagen, aggrecan and chondroitin-6-sulfotransferase. • Collagen gels were capable of inducing cell proliferation, but they did not show substantial ECM synthesis or gene expression. • Properties were lowest in agarose, alginate and fibrin gels.	Gruber et al., 1997

Table 2.2: *Continued.* Biodegradable polymer scaffold based tissue engineering for IVD regeneration *Continues.*

Scaffold	Cell	Method	Findings	Reference
Collagen type I scaffolds	NP cells of human lumber IVDs	The scaffolds consisted of highly dense (0.5%–12%) type I collagen matrices, prepared by plastic compression.	• The efficacy of the collagen sponge is attributed to its porous structure that may have facilitated increased ECM production and diffusion of molecules. • Gamma sterilization of the scaffolds increased the shear moduli but also resulted in more brittle behavior and a reduced swelling capacity, dense collagen is a promising candidate for tissue engineering of the NP.	Bron et al., 2009
Three-dimensional (3D) constructs in which cells were seeded on polyester fiber meshes and encapsulated in calcium crosslinked alginate	Outer and inner annulus fibrosus and nucleus pulposus	Three-dimensional (3D) constructs were cultured for 14 days *In vitro* and evaluated histologically and quantitatively for gene expression and production of types I and II collagen and proteoglycans.	• Culture environment may have a greater impact on cellular behavior.	Chou et al., 2009
Hybrid scaffolds of PLGA (poly (lactic-co-glycolic) acid), SIS (small intestinal submucosa) and DBP (demineralized bone particles)	AF cells of rabbit	PLGA, PLGA/SIS(20%), PLGA/DBP(20%) and PLGA/SIS (10%)/DBP(10%) scaffold were manufactured by solvent casting/salt leaching method.	• Using DBP(especially 20 wt% of DBP) in terms of scaffold fabrication for bio-disc with IVD cells helps growth of disc cells maintained their phenotypes.	Ha et al., 2008
PLGA-Fibrin hybrid scaffold	Rabbit chondrocytes cells	PLGA scaffolds were soaked in chondrocytes-fibrin suspension (1×10^6cells/scaffold) and polymerized by dropping thrombin-calcium chloride (CaCl2) solution.	• Fibrin/PLGA promotes early *in vitro* chondrogenesis of rabbit articular chondrocytes. • Serve as potential cell delivery vehicle.	Sha'ban et al., 2008
Atelocollagen type II	Bovine nucleus pulposus cell	Bovine nucleus pulposus cells that were seeded within non-cross-linked and enzymatically cross-linked, atelocollagen type II based scaffolds containing varying concentrations of aggrecan and hyaluronan.	• Cross-linking atelocollagen type II based scaffolds did not cause any negative effects on cell viability or cell proliferation over the 7-day culture period. • The cross-linked scaffolds retained the highest proteoglycan synthesis rate and the lowest elution of sulfated glycosaminoglycan into the surrounding medium. • Cross-linked scaffolds provided a more stable structure for the cells compared to the non-cross-linked scaffolds.	Halloran et al., 2008
PLGA scaffold	NP cells from disc of adult female rabbit	PLGA scaffolds were prepared by solvent casting/salt-leaching and cells were seeded in prepared PLGA scaffold and cultured in	• Compression strength of scaffold decreased with increasing porogen size.	Hong et al., 2009

Table 2.2: *Continued.* Biodegradable polymer scaffold based tissue engineering for IVD regeneration.

HA loaded PLGA scaffolds	Chondrocytes	• HA loaded PLGA scaffolds were prepared by a emulsion freeze-drying method. The chondrocytes were seeded on the HA-PLGA scaffolds • The chondrocytes were seeded on the HA-PLGA scaffolds and measured by MTT assay. Morphological observation, histology, biological assay for collagen and sGAG, and PCR were performed	• Scaffolds containing HA were higher cell viability then only PLGA scaffolds • HA/PLGA scaffold may serve as a potential cell delivery vehicle and a structural basis for *in vitro* tissue engineered articular cartilage	Jung *et al.*, 2008
Nanofibrous scaffold (NFS) enveloping a hyaluronic acid (HA) hydrogel center	Adult human mesenchymal stem cells (MSCs)	• Biphasic construct made of novel biomaterial Amalgam that consisted of electrospun, biodegradable nanofibrous scaffold (NFS) enveloping a hyaluronic acid (HA) hydrogel center	• Time-dependent development of chondrocytic phenotype of the seeded cells. • The cells also maintain the microarchitecture of a native IVD	Nesti *et al.*, 2008

acid) (PLGA) scaffolds with macro-porous structure are a very common model for tissue engineered constructs (Gunatillake and Adhikari, 2003). PLGA scaffolds supported NP cell proliferation and glycosaminoglycan (GAG) production, although at a lower level than when the NP cells were dispersed in a fibrin gel seeded within the PLGA macro-structure (Hong et al., 2009; Sha'ban et al., 2008). Similar to what had already been observed for chondrocytes in macro-porous polyurethane scaffolds and fibrin/polyurethane scaffold combinations (Lee et al., 2005), these results can be attributed to the lower cell seeding efficiency and the 2D environment experienced by the seeded NP cells, which is not optimal for NP cell culture as compared to the true 3D structure provided by the fibrin gel. This partly explains the focus of the past and present research on hydrated biopolymer structures.

Recent work on synthetic polymers tends to emulate biopolymer hydrogels and mimic some of the structural features of the NP extracellular matrix. A fully synthetic thermoresponsive PLGA-poly(N-isopropylacrylamide) hydrogel and its potential as an injectable carrier for NP cell delivery have recently been described (Fraylich et al., 2010). Synthetic polymeric nano-fiber based scaffolds mimicking ECM nano-structure for the control of IVD cell behavior have also emerged as a prospective approach toward biofunctional 3D scaffolds for annulus fibrosus, but also nucleus pulposus repair (Nerurkar et al., 2009; Yang et al., 2009). Nevertheless, compared to biopolymers, synthetic polymers and new synthetic cell delivery systems for NP regeneration have barely been investigated either *in vitro* or *in vivo*. Most of the biomaterials reported to date have first been employed and developed to improve understanding and manipulation of the NP cells *in vitro* and secondly for the delivery of relevant cell sources (NP cells and MSCs) *in vivo*.

2.7 POLYMERIC HYDROGELS FOR *IN VITRO* NUCLEUS PULPOSUS REGENERATION STUDIES

A wide variety of hydrogels has been investigated in NP regeneration and is summarized in Table 2.3. Hydrogels made from alginate, a naturally derived polysaccharide originating from brown algae, are probably the most extensively investigated matrices for the encapsulation of IVD cells *in vitro* (Chiba et al., 1997; Yang et al., 2008; Chou and Nicoll, 2009; Kim et al., 2009). Nucleus pulposus cells are routinely cultured by encapsulation in alginates (Baer et al., 2001; Gruber et al., 1997b; Maldonado and Oegema, 1992; Melrose et al., 2001; Wang et al., 2001). The predominant method of alginate gelation is through ionic cross-linking, achieved via diffusion of divalent cations (e.g., calcium) to carboxylic acid moieties on the polymer, resulting in a crosslinked network. Photocrosslinked alginate has also been used for NP cellular encapsulation and synthesized with tunable material properties that may be adequate for IVD regeneration (Chou and Nicoll, 2009). NP cells of different sources (e.g., rabbit, porcine, bovine, and human) have been encapsulated into alginate beads and cultured *in vitro* (Chiba et al., 1997; Baer et al., 2001; Zeiter et al., 2009; Sakai et al., 2006). NP cell encapsulation in alginate hydrogel promotes a rounded, chondrocyte-like morphology, in contrast to the elongated, fibroblast-like morphology seen in monolayer cultures. They usually conserve their characteristic phenotype with expression of type II collagen. Therefore, alginate but also agarose gels, which provide a very similar environment and have extensively been used for the study of NP cells in 3D culture systems and for the assessment of biological and general environmental influences on NP cell behavior. The effects of drugs and biologically relevant molecules like the bone morphogenetic protein 2 (BMP-2) on NP cells have been reported using alginate 3D culture systems (Lee et al., 2010; Kim et al., 2009). NP cells positively respond to mechanical compression and to a lesser extent to frequency change when seeded in alginate (Korecki et al., 2009). However, initially produced stable alginate gels and their mechanical integrity have been found to decrease over time, possibly due to a loss of ions through diffusion (Baer et al., 2001) or depletion by the encapsulated cells. The effects of physical stimuli like ultrasound, hypoxia, or hydrostatic pressure on NP cell behavior have also been tested *in vitro* using alginate beads (Kobayashi et al., 2009; Erwin et al., 2009; Rastogi et al., 2009; Le Maitre et al., 2008).

Overall, alginate hydrogels provide a useful and well-established 3D matrix for the *in vitro* study of NP cell behavior under controlled experimental conditions; however, most of the data collected using alginate hydrogels are difficult to compare due to the different sources of alginate and gelation procedures used that can significantly influence the properties of the 3D matrix (e.g., variations in stiffness) (Augst et al., 2006). Nonetheless, the potential of different progenitor cell sources for the repair of NP has also been evaluated in alginate beads *in vitro* (Xie et al., 2009; Gaetani et al., 2008). However, since the survival of alternative therapeutic cells like MSCs in alginate gel may be impaired due to the lack of specific binding sites, its application for use in NP cell therapy is critical (Zeiter et al., 2009). Early chondrogenic induction and conditioning of MSCs can improve their survival, indicating that pre-conditioning of cells or combination of growth factors and MSCs in an alginate delivery system could be successful for NP regeneration therapy.

Substrate attachment sites are necessary for growth, differentiation, replication, and metabolic activity of many cell types (e.g., MSCs) in culture. For example, cells can attach to type I collagen, which can serve as tissue regeneration scaffold for a number of cellular constructs. Chondrocytes, for instance, can retain their phenotype and cellular activity when cultured on collagen (Toolan et al., 1996). To date, type I collagen has mostly been considered for AF repair in combination or not with other polymers (Bowles et al., 2010). Human IVD cells seeded onto collagen sponge showed improved cell proliferation and lower sulfated glycosaminoglycan content after 14 days *in vitro* compared to agarose gel (Gruber et al., 2004). Similarly, atelocollagen hydrogels of different concentration, 0.3 and 3% w/v, have been considered for the culture of human NP cells. Atelocollagen performed best in terms of DNA increase, but worse or similarly to alginate hydrogel in terms of proteoglycan synthesis and accumulation (Sakai et al., 2006), indicating that the properties (e.g., surface charge and degradation rate) of the collagen hydrogels are less favorable than the alginate gel for NP matrix retention. The main advantages of atelocollagen hydrogels, in addition to providing a better structure for cell attachment and proliferation than alginate, are related to their safety and their biomedical market availability for future applications. The advantageous properties of collagen for supporting tissue growth have been used in conjunction with the superior mechanical properties of other natural or synthetic biodegradable polymer systems to produce hybrid tissue scaffolds

Table 2.3: Hydrogel based tissue engineering for IVD regeneration. *Continues.*

Hydrogels	Methods	Outcomes	References
Poly(2-hydroxyethylmethacrylate)/ polycaprolactone (PHEMA/PCL)	• These interpolymer network of PHEMA/PCL was used in intervertebral disc prostheses.	Composite PHEMA/PCL hydrogels showed compression properties similar to those expressed by canine intervertebral discs in different spinal locations.	Ambrosio et al., 1998
Acrylic acid/ acrylamide	• The partition and diffusion characteristics of an acrylic acid/acrylamide hydrogel, copolymerized in the pores of a polyurethane foam with sodium and chloride ions, were studied.	Comparison of the hydrogel foam with cartilage and intervertebral disc shows considerable similarities.	Lanir et al., 1998
Copolymer of poly(vinyl alcohol) (PVA) and poly(vinyl pyrrolidone) (PVP)	• Prosthetic disc nucleus device consists of two hydrogel pellets, each enclosed in a woven polyethylene jacket.	The original intact disc height was restored after implantation of the device.	Wilke et al., 2001; Thomas et al.,2004
Poly (vinyl alcohol)	• PVA implants were inserted into discectomy defects created in the L3-L4 or L4-L5 intervertebral disc in 20 male baboons.	The PVA implants were well tolerated over 24 months *in vivo*. No evidence of device-related pathology in the adjacent disc tissue, spinal cord, or remote tissues.	Allen et al.,2004
Semicrystalline PVA hydrogel elastomers	• Aqueous PVA solution froze at -20 degrees C for 6-12 h, and then thawed at room temperature for 1-2 h. The same process was repeated 1-3 times. After the specimen was dehydrated in vacuum, a kind of artificial disc nucleus materials was formed.	An artificial disc nucleus materials was successfully developed and characterized.	Gu et al., 2004
Copolymers of 2-(4'-iodobenzoyl)-oxo-ethyl methacrylate (4IEMA) and a hydrophilic building block (either N-vinyl-2-pyrrolidinone (NVP) or 2-hydroxyethyl methacrylate (HEMA)	• 4 copolymers of NVP/4IEMA and 4 copolymers of HEMA/4IEMA in different compositions (5, 10, 15 and 20 mol% 4IEMA) were prepared.	Hydrogels with 5 mol% 4IEMA appear to meet all criteria; they are non-cytotoxic, have adequate physical-mechanical properties and feature sufficient radiopacity in a realistic model.	Boelen et al., 2005
Hyaluronic acid (HA), PEG-g-chitosan, Agarose, Gelatin, Alginate	• Injectable NP replacements or tissue engineering scaffolds were prepared and	The synthetic HA-based hydrogels approximated NP well and may serve as	Cloyd et al., 2007

Table 2.3: *Continued.* Hydrogel based tissue engineering for IVD regeneration. *Continues.*

Hydrogels	Methods	Outcomes	References
Poly(2-hydroxyethylmethacrylate)/ polycaprolactone (PHEMA/PCL)	• These interpolymer network of PHEMA/PCL was used in intervertebral disc prostheses.	• Composite PHEMA/PCL hydrogels showed compression properties similar to those expressed by canine intervertebral discs in different spinal locations.	Ambrosio *et al.*, 1998
Acrylic acid/ acrylamide	• The partition and diffusion characteristics of an acrylic acid/acrylamide hydrogel, copolymerized in the pores of a polyurethane foam with sodium and chloride ions, were studied.	• Comparison of the hydrogel foam with cartilage and intervertebral disc shows considerable similarities.	Lanir *et al.*, 1998
Copolymer of poly(vinyl alcohol) (PVA) and poly(vinyl pyrrolidone) (PVP)	• Prosthetic disc nucleus device consists of two hydrogel pellets, each enclosed in a woven polyethylene jacket	• The original intact disc height was restored after implantation of the device.	Wilke *et al.*, 2001; Thomas *et al.*,2004
Poly (vinyl alcohol)	• PVA implants were inserted into discectomy defects created in the L3-L4 or L4-L5 intervertebral disc in 20 male baboons.	• The PVA implants were well tolerated over 24 months *in vivo*. • No evidence of device-related pathology in the adjacent disc tissue, spinal cord, or remote tissues.	Allen *et al.*,2004
Semicrystalline PVA hydrogel elastomers	• Aqueous PVA solution froze at -20 degrees C for 6-12 h, and then thawed at room temperature for 1-2 h. The same process was repeated 1-3 times. After the specimen was dehydrated in vacuum, a kind of artificial disc nucleus materials was formed.	• An artificial disc nucleus materials was successfully developed and characterized.	Gu *et al.*, 2004
Copolymers of 2-(4'-iodobenzoyl)-oxo-ethyl methacrylate (4IEMA) and a hydrophilic building block (either N-vinyl-2-pyrrolidinone (NVP) or 2-hydroxyethyl methacrylate (HEMA)	• 4 copolymers of NVP/4IEMA and 4 copolymers of HEMA/4IEMA in different compositions (5, 10, 15 and 20 mol% 4IEMA) were prepared.	• Hydrogels with 5 mol% 4IEMA appear to meet all criteria; they are non-cytotoxic, have adequate physical-mechanical properties and feature sufficient radiopacity in a realistic model.	Boelen *et al.*, 2005
Hyaluronic acid (HA), PEG-g-chitosan, Agarose, Gelatin, Alginate	• Injectable NP replacements or tissue engineering scaffolds were prepared and	• The synthetic HA-based hydrogels approximated NP well and may serve as	Cloyd *et al.*, 2007

Table 2.3: *Continued.* Hydrogel based tissue engineering for IVD regeneration. *Continues.*

Hydrogels	Methods	Outcomes	References
Poly(2-hydroxyethylmethacrylate)/polycaprolactone (PHEMA/PCL)	• These interpolymer network of PHEMA/PCL was used in intervertebral disc prosthesis.	• Composite PHEMA/PCL hydrogels showed compression properties similar to those expressed by canine intervertebral discs in different spinal locations.	Ambrosio et al., 1998
Acrylic acid/ acrylamide	• The partition and diffusion characteristics of an acrylic acid/acrylamide hydrogel, copolymerized in the pores of a polyurethane foam with sodium and chloride ions, were studied.	• Comparison of the hydrogel foam with cartilage and intervertebral disc shows considerable similarities.	Lanir et al., 1998
Copolymer of poly(vinyl alcohol) (PVA) and poly(vinyl pyrrolidone) (PVP)	• Prosthetic disc nucleus device consists of two hydrogel pellets, each enclosed in a woven polyethylene jacket.	• The original intact disc height was restored after implantation of the device.	Wilke et al., 2001; Thomas et al.,2004
Poly (vinyl alcohol)	• PVA implants were inserted into discectomy defects created in the L3-L4 or L4-L5 intervertebral disc in 20 male baboons.	• The PVA implants were well tolerated over 24 months *in vivo*. No evidence of device-related pathology in the adjacent disc tissue, spinal cord, or remote tissues.	Allen et al.,2004
Semicrystalline PVA hydrogel elastomers	• Aqueous PVA solution froze at -20 degrees C for 6-12 h, and then thawed at room temperature for 1-2 h. The same process was repeated 1-3 times. After the specimen was dehydrated in vacuum, a kind of artificial disc nucleus materials was formed.	• An artificial disc nucleus materials was successfully developed and characterized.	Gu et al., 2004
Copolymers of 2-(4'-iodobenzoyl)-oxo-ethyl methacrylate (4IEMA) and a hydrophilic building block (either N-vinyl-2-pyrrolidinone (NVP) or 2-hydroxyethyl methacrylate (HEMA)	• 4 copolymers of NVP/4IEMA and 4 copolymers of HEMA/4IEMA in different compositions (5, 10, 15 and 20 mol% 4IEMA) were prepared.	• Hydrogels with 5 mol% 4IEMA appear to meet all criteria; they are non-cytotoxic, have adequate physical-mechanical properties and feature sufficient radiopacity in a realistic model.	Boelen et al., 2005
Hyaluronic acid (HA), PEG-g-chitosan, Agarose, Gelatin, Alginate	• Injectable NP replacements or tissue engineering scaffolds were prepared and	• The synthetic HA-based hydrogels approximated NP well and may serve as	Cloyd et al., 2007

Table 2.3: *Continued.* Hydrogel based tissue engineering for IVD regeneration.

Hydrogels	Methods	Outcomes	References
Poly(2-hydroxyethylmethacrylate)/ polycaprolactone (PHEMA/PCL)	• These interpolymer network of PHEMA/PCL was used in intervertebral disc prostheses.	• Composite PHEMA/PCL hydrogels showed compression properties similar to those expressed by canine intervertebral discs in different spinal locations.	Ambrosio et al., 1998
Acrylic acid/ acrylamide	• The partition and diffusion characteristics of an acrylic acid/acrylamide hydrogel, copolymerized in the pores of a polyurethane foam with sodium and chloride ions, were studied.	• Comparison of the hydrogel foam with cartilage and intervertebral disc shows considerable similarities.	Lanir et al., 1998
Copolymer of poly(vinyl alcohol) (PVA) and poly(vinyl pyrrolidone) (PVP)	• Prosthetic disc nucleus device consists of two hydrogel pellets, each enclosed in a woven polyethylene jacket.	• The original intact disc height was restored after implantation of the device.	Wilke et al., 2001; Thomas et al.,2004
Poly (vinyl alcohol)	• PVA implants were inserted into discectomy defects created in the L3-L4 or L4-L5 intervertebral disc in 20 male baboons.	• The PVA implants were well tolerated over 24 months *in vivo*. • No evidence of device-related pathology in the adjacent disc tissue, spinal cord, or remote tissues.	Allen et al.,2004
Semicrystalline PVA hydrogel elastomers	• Aqueous PVA solution froze at -20 degrees C for 6-12 h, and then thawed at room temperature for 1-2 h. The same process was repeated 1-3 times. After the specimen was dehydrated in vacuum, a kind of artificial disc nucleus materials was formed.	• An artificial disc nucleus materials was successfully developed and characterized.	Gu et al., 2004
Copolymers of 2-(4'-iodobenzoyl)-oxo-ethyl methacrylate (4IEMA) and a hydrophilic building block (either N-vinyl-2-pyrrolidinone (NVP) or 2-hydroxyethyl methacrylate (HEMA)	• 4 copolymers of NVP/4IEMA and 4 copolymers of HEMA/4IEMA in different compositions (5, 10, 15 and 20 mol% 4IEMA) were prepared.	• Hydrogels with 5 mol% 4IEMA appear to meet all criteria; they are non-cytotoxic, have adequate physical-mechanical properties and feature sufficient radiopacity in a realistic model.	Boelen et al., 2005
Hyaluronic acid (HA), PEG-g-chitosan, Agarose, Gelatin, Alginate	• Injectable NP replacements or tissue engineering scaffolds were prepared and	• The synthetic HA-based hydrogels approximated NP well and may serve as	Cloyd et al., 2007

for IVD regeneration (Rong et al., 2002; Alini et al., 2003; Sato et al., 2003). These hybrid systems usually show superior cell adhesion, interaction, and proliferation due to the collagen as compared to the polymer alone system. Gelatin, a denatured collagen is also commonly used for pharmaceutical and medical applications because of its biodegradability (Ikada and Tabata, 1998; Kawai et al., 2000; Yamamoto et al., 2001; Balakrishnan and Jayakrishnan, 2005) and biocompatibility in physiological environments (Kuijpers et al., 2000a; Yao et al., 2004). These characteristics have contributed to gelatin's proven record of safety as a plasma expander, as an ingredient in drug formulations, and as a sealant for vascular prostheses (Kuijpers et al., 2000b). Recently, a tissue-engineered construct made of gelatine and demineralised bone matrix seeded with porcine NP cells in polylactide scaffold was cultured *in vitro* (Brown et al., 2005). Fibroblast-like cell morphology was reported for the NP cells with up-regulation of type I and II collagen gene expression, indicating that not all biopolymer hydrogels may be adequate for NP cell culture *in vitro* and that controlling the biological cell response to the artificial matrix environment is a complex task.

Glycosaminoglycans, hyaluronan and mixtures have been investigated for IVD tissue engineering applications. Glycosaminoglycans are polysaccharides which occur ubiquitously within the extracellular matrix (ECM) of most animals and are important components of the NP tissue together with hyaluronan. The predominant types of GAGs attached to naturally occurring core proteins of proteoglycans include chondroitin sulfate, dermatan sulfate, keratan sulfate, and heparan sulfate (Heinegard and Paulsson, 1980; Naeme and Barry, 1993). NP, AF, and mesenchymal stem cells (MSCs) have been cultured in hyaluronan in the presence of growth factors (Crevensten et al., 2004; Nesti et al., 2008). Recently, Jung and coworkers (Jung et al., 2008) prepared and characterized hyaluronic acid loaded PLGA scaffolds for chondrocyte cells using an emulsion freeze drying method. They confirmed the higher cell viability in hyaluronan loaded PLGA scaffolds than in the PLGA only scaffolds and its potential as a cell delivery vehicle and a structural basis for *in vitro* tissue engineered articular cartilage. Injectable hyaluronic acid hydrogels are of major interest and several technologies are currently being developed for the optimal delivery of cells in degenerating NP (Cloyd et al., 2007; Su et al., 2010; Mortisen et al., 2010). Elisseeff and co-workers have developed a photopolymerization method to successfully encapsulate chondrocytes in synthetic poly(ethylene oxide)-based (PEO) hydrogels for tissue engineering applications (Elisseeff et al., 1999). This technique was modified by Bryant and coworkers (Bryant et al., 2003) to incorporate degradable lactic acid units into poly(ethylene glycol)-based (PEG) hydrogels and enhance the spatial distribution of ECM components in these otherwise non-degradable polymers. Then the photopolymerization was employed to create polysaccharide-based hydrogels using alginate and hyaluronic acid macromers modified with functional methacrylate groups (Smeds et al., 2001; Burdick et al., 2005; Nettles et al., 2004). Injectable hyaluronan matrix for cell delivery is able to maintain the NP phenotype (type II collagen and aggrecan) and provide an enzymatically degradable matrix that can be replaced by the regenerated tissue. Interestingly, hyaluronan molecules have been shown to influence IVD cells similarly as transforming growth factor-beta 3 (TGF-β3) towards an NP phenotype and NP matrix formation *in vitro* (Haberstroh et al., 2009). Although hyaluronic acid-based hydrogels have shown promising results, these biomaterials are often derived from an animal source, which presents the risk of batch-to-batch variations and the need for additional purification steps to reduce the possibility of stimulating an immune response upon implantation. Nevertheless, the results observed for hyaluronic acid-based hydrogels indicate the potential of a polysaccharide-based system. Carboxymethylcellulose (CMC, a water-soluble derivative of cellulose), a biocompatible, low cost, FDA-approved biomaterial that is commercially available in high-purity forms, has been considered as an alternative option to hyaluronan (Boccaccini and Blaker, 2005), and photocrosslinked CMC hydrogels were reported for NP cell encapsulation (Reza and Nicoll, 2010). Changing the CMC gel composition and concentration influenced the mechanical properties of bovine NP cells/CMC tissue-engineered constructs cultured for 7 days. Interplay between the stiffness and the degradation of the hydrogel via hydrolysis and the building up and lying down of the extracellular matrix was suggested to be a crucial parameter for the tailoring of NP tissue-engineered constructs.

Recently, we developed an ECM mimicking hydrogel for NP regeneration based on hyaluronan and atelocollagen type II (Halloran et al., 2008). In this work, the behaviour of bovine NP cells was investigated within non-cross-linked and enzymatically cross-linked atelocollagen type II based injectable scaffolds containing varying concentrations of aggrecan and hyaluronan. Cross-linking atelocollagen type II did not cause any negative effects on cell viability or cell proliferation over the 7-day culture period. The cross-linked scaffolds retained the highest proteoglycan synthesis rate and the lowest elution of sulfated glycosaminoglycan into the surrounding medium. The results of this study indicate that the enzymatically cross-linked, composite collagen-hyaluronan scaffold shows great potential for developing an injectable cell-seeded scaffold for nucleus pulposus treatment in degenerative intervertebral discs (Halloran et al., 2008). This may again support the importance of the relationship between the rates of new matrix production and hydrogel degradation. Others have reported on the preparation of a non-injectable collagen type II/hyaluronan/chondroitin-6-sulfate scaffold using a lyophilisation process and chemical binding of the chondroitin-6-sulfate on the surface of the scaffold (Huang et al., 2010). The reported biomaterial is less relevant in terms of application as it does not permit the delivery of cells via injection; nonetheless, the authors showed an increase in collagen type II and proteoglycan gene expression, decrease in collagen type I gene expression and increase in GAG production by seeded NP cells in the scaffold compared to monolayer culture.

The present knowledge about biomaterials for NP cell encapsulation indicates the importance of providing a 3D environment and the necessity for biological recognition of the material surface for cell attachment, survival, proliferation and matrix production. Other important scaffold properties capable of influencing NP cell behavior, like mechanics, permeability and degradation, have only partially been studied and not yet optimized *in vitro* explaining the generally poor rationale for biomaterial selection in *in vivo* studies compared to more established bone tissue engineering approaches. Cloyd et al. (2007) have recently compared the mechanical properties of injectable hydrogels based on alginate, hyaluronan, or poly(ethylene glycol) for nucleus repair with human nucleus pulposus tissue. Mechanical properties of NP (linear modulus of 5.39 kPa, Poisson's ratio 0.62, relaxation 65.77% measured in an unconfined compression mode) were approached by hyaluronan cross-link compositions, but their combination with cells has not yet been reported (Cloyd et al., 2007).

2.8 POLYMERIC HYDROGELS FOR *IN VIVO* NUCLEUS PULPOSUS REGENERATION

Injectable polymeric hydrogels for NP replacement and the delivery of cells, mainly NP cells or MSCs, in the degenerating IVD are the most relevant materials developed for providing minimal lesion to the AF tissue during surgery and avoiding complete replacement of the IVD. Ruan et al. (2010) though inserted NP cells that were cultured on PLGA scaffold before implantation in a beagle dog model. Living NP cells seeded onto the PLGA scaffold were found after 8 weeks *in vivo*, and it was suggested that the PLGA tissue-engineered construct could prevent or delay the degenerative process of the IVD.

Like the synthetic PLGA scaffold, injections of acellular hyaluronan gels into NP of primate, pig, and rat models after nucleotomy have also demonstrated to decrease the degeneration of damaged discs (Pfeiffer et al., 2003; Revell et al., 2007; Crevensten et al., 2004). Similar or improved delaying of the *in vivo* degeneration of the disc has been reported for the combination of hyaluronan gel and MSCs. For example, hyaluronan was used to grow adult rat bone marrow MSCs and inject them into healthy rat discs. The injection site in the AF was closed after 28 days and tissue resembling native NP tissue had formed. A substantial increase in number of labeled MSCs and disc height at day 28 was also reported (Crevensten et al., 2004). These findings suggest the potential of hyaluronan as medium to grow relevant cells *in vitro* in the form of scaffolds and injectable biomaterials for IVD regeneration. Nonetheless, the extent of the cell therapy success compared to a pure biomaterial approach is not clear yet due to differences in *in vivo* models and outcome parameters (e.g., MSC survival, IVD height) analyzed.

Other gels that are under scrutiny and have been tested *in vivo* combined with MSCs in disc degeneration models include atelocollagen (Sakai et al., 2003, 2006), synthetic peptide hydrogel (Henriksson et al., 2009), fibrin and hyaluronan/fibrin gels (Bertram et al., 2005; Stern et al., 2000), modified alginate (Leone et al., 2008), and chitosan (Liu et al., 2009). Human MSCs injected with or without a peptide hydrogel matrix survived for up to 6 months in a xenogeneic porcine model, although all groups of IVDs showed signs of degeneration (Henriksson et al., 2009).

So far, no clear advantage of one biomaterial over another or of the use of a hydrogel to improve the outcome of the therapy could be demonstrated. Therefore, biomaterial screening and optimization for NP regeneration therapy is still necessary, not only to test the ability of the biomaterials to perform at each stage of the therapy envisioned (e.g., injectability, spatial, and temporal gelling, mechanics), but also to develop modeling and experimental tools. In this context, *ex vivo* implant whole IVD culture bioreactors mimicking *in vivo* conditions (e.g., nutrition and mechanics) are valuable to optimize biomaterials and reduce animal studies (Yao et al., 2006; Jünger et al., 2009).

2.9   FUTURE PERSPECTIVES

Although research into NP regeneration is still relatively in its infancy, efforts to regenerate and replace the tissues of the intervertebral disc have virtually exploded over the last two decades and shown much promise for providing alternative treatment options to arthrodesis, TDR, or NPR surgeries as long as patients presenting with discogenic LBP are diagnosed at an early enough stage. It is imperative that this degenerative process becomes completely understood because most of the therapies are based on biological manipulation of native degenerative tissue. Tissue engineering strategies must take into account the unfavorable environment for cell activity that is present in degenerating IVDs. It may not suffice to implant cell-seeded constructs into degenerated tissue that is already strained by nutrient deficiency, low oxygen levels, high lactic acid concentrations, up-regulated catabolic gene expression levels, and, possibly, altered biomechanical stresses (Bibby et al., 2005). Therefore, the choice of a scaffold material is a critical factor in protecting the implanted cells from this harsh environment (O'Halloran et al., 2005; O'Halloran and Pandit, 2007) and may well depend on the combination of the degree of NP damage and the therapy envisioned.

The choice of scaffold composition has been shown to play an important role in NP cell behavior (Table 2.2). This temporary local environment of the cells could provide certain molecular signals but also drugs or genes that promote expression of the required cell phenotype. In particular, with the use of adult MSCs, the scaffold could play a significant role in inducing cell differentiation towards NP cells. Although promising results have been observed, and much information gained from *in vitro* and *in vivo* trials using NP cells, it will not be practical in the clinical setting to acquire autologous disc cells for implantation. Therefore, it is likely that most cell-based therapies will involve the use of autologous adult MSCs.

Recently, drug and gene therapy has also been considered as an alternative therapeutic approach for delivery of biologics in IVD regeneration (Table 2.4) (Nishida et al., 1999; Wehling et al., 1997; Yoon, S., 2005; Yoon et al., 2004; Le Maitre et al., 2006). Biomaterials for controlled delivery of plasmid DNA can provide an essential tool to promote localized transgene expression, which can be employed to direct cellular processes for numerous tissue engineering applications. The delivery of genes *in situ* for NP repair has already been proposed using chitosan hydrogel (Alini et al., 2002; Di Martino et al., 2005b), but so far no studies have reported on the development and use of a biomaterial based system for the spatial and temporal delivery of drugs and/or genes in degenerative NP in combination or not with cells.

Only a few tissue-engineering solutions have been proposed to integrate AF and NP tissues in the repair process, and additional work to promote integration among native, neogenerated, and implanted tissues will be critical to restoring IVD function. Temporary recapitulation of the NP mechanical properties with a material based therapy is most probably a key issue to address in order to successfully stabilize the organ and permit healing of a functional tissue. Thus, advances in IVD cellular and molecular biology are needed to enable the identification of novel therapeutic targets, to select for classes of biomaterials suitable for cell-biomaterial interactions and ECM deposition, and to suggest appropriate drug and gene delivery strategies.

Table 2.4: List of biomaterials used in biological(s) based therapies in IVD regeneration.

Materials	Signals	Tissues/ Cells used	Sources	References
Alginate	TGF-β1	IVD cells	Humans; normal, patient, young and old	Gruber et al., 1997
Matrigel™ [laminin (a major component), collagen IV, heparan sulfate proteoglycans]	bFGF, EGF, ILGF-1, PDGF TGF-β1, NGF	AF cells	Human (patients with DDD)	Desai et al., 1999
Collagen type I	TGF-β, bFGF, IGF-1	NP and AF cells	Two-three year old bovine steers	Alini et al., 2003
Hyaluronan	TGF-β, bFGF, IGF-1	NP and AF cells	Two-three year old bovine steers	Alini et al., 2003
Alginate	rhBMP-2	NP and transition zone cells	Human IVDs; grade III and IV degeneration	Kim et al., 2003
	rhOP-1	NP and inner AF cells	Adolescent rabbit IVDs (majority of cells in NP are notochordal cells)	Masuda et al., 2003
	OP-1	NP and AF cells	Rabbit IVDs	An et al., 2003a
	GDF-5	IVD tissue	Naturally occurring mutant GDF-5 deficient mice	Li et al., 2004
	OP-1	NP and AF cells	Rabbits (containing notochordal cells)	Takegami et al., 2005
	Platelets rich plasma (PRP) based growth factor	NP and AF cells	Porcine	Akeda et al., 2006
Gelatin	Platelets rich plasma (PRP) based growth factor	-	-	Sawamura et al., 2009; Nagae et al., 2007

2.10   ACKNOWLEDGEMENTS

We thank Paula Walsh (graphical artist) for her contribution in graphical illustration and Dr. Wenxin Wang for technical discussion. The authors would also like to acknowledge Science Foundation Ireland for their support under the research under Grant No. 07/SRC/B1163 and 07/RFP/ENMF482.

REFERENCES

Adams, M., Freeman, B., Morrison, H., Nelson, I., Dolan, P. (2000). Mechanical initiation of intervertebral disc degeneration. *Spine* 25, 1625–36. DOI: 10.1097/00007632-200007010-00005 48

Adler, J.H., Schoenbaum, M., Silberberg, R. (1983). Early onset of disk degeneration and spondylosis in sand rats (Psammomys obesus). *Vet Pathol* 20(1), 13–22. 50

Aguiar, D.J., Johnson, S.L., Oegema, J., Theodore, R. (1999). Notochordal cells interact with nucleus pulposus cells: Regulation of proteoglycan synthesis. *Exp Cell Res* 246(1), 129–37. DOI: 10.1006/excr.1998.4287 52

Akeda, K., An, H. S., Pichika, R., Attawia, M., Thonar, E., Lenz, M. E., Uchida, A., and Masuda, K. (2006). Platelet-rich plasma (PRP) stimulates the extracellular matrix metabolism of porcine nucleus pulposus and anulus fi brosus cells cultured in alginate beads. *Spine* 31(9), 959–66. DOI: 10.1097/01.brs.0000214942.78119.24

Alini, M., Markovic, P., Aebi, M., Spiro, R., Roughley, P. (2003) The potential and limitations of a cell-seeded collagen/hyaluronan scaffold to engineer an intervertebral disc-like matrix. *Spine* 28, 446. 65

Alini, M., Roughley, P. J., Antoniou, J., Stoll, T., Aebi, M. (2002). A biological approach to treating disc degeneration: not for today, but maybe for tomorrow. *Eur Spine J* 11 (Suppl 2), S215–20. DOI: 10.1007/s00586-002-0485-8 68

Allen, M. J., Schoonmaker, J. E., Bauer, T. W., Williams, P. F., Higham, P. A., Yuan, H. A. (2004). Preclinical evaluation of a poly (vinyl alcohol) hydrogel implant as a replacement for the nucleus pulposus. *Spine* 29(5), 515–23. DOI: 10.1097/01.BRS.0000113871.67305.38 54

Ambrosio, L., De Santis, R., Nicolais, L. (1998). Composite hydrogels for implants. *Proc Inst Mech Eng H* 212(2), 93–9. DOI: 10.1243/0954411981533863

Aminian, O., Mehrdad, R., Berenji, M. (2003). Prevalence of low back pain and disability in bank office workers in western Tehran. The Global Occupational Health Network 6. 43

An, H., Boden, S.D., Kang, J., Sandhu, H.S., Abdu, W., Weinstein, J. (2003a). Summary statement: emerging techniques for treatment of degenerative lumbar disc disease. *Spine* 28 (15 Suppl.), S24. 43, 49

An, H.S., Thonar, E.J., Masuda, K. (2003b). Biological repair of intervertebral disc. Spine 28 (15 Suppl.), S86. 51

Anderson, J. Epidemiological aspects of back pain. (1986). *J Soc Occup Med* 36, 90–4. 43

Annabi, N., Mithieux, S. M., Boughton, E. A., Ruys, A. J., Weiss, A. S., Dehghani, F. (2009). Synthesis of highly porous crosslinked elastin hydrogels and their interaction with fibroblasts *in vitro*. *Biomaterials* 30, 4550–7. DOI: 10.1016/j.biomaterials.2009.05.014

Annabi, N., Mithieux, S. M., Weiss, A. S., Dehghani, F. (2010). Cross-linked open-pore elastic hydrogels based on tropoelastin, elastin and high pressure CO_2. *Biomaterials* 31(7), 1655–65. DOI: 10.1016/j.biomaterials.2009.11.051

Antoniou, J., Steffen, T., Nelson, F., Winterbottom, N., Hollander, A.P., Poole, R.A., Aebi, M., Alini, M. (1996). The human lumbar intervertebral disc: evidence for changes in the biosynthesis and denaturation of the extracellular matrix with growth, maturation, ageing and degeneration. *J Clin Invest* 98, 996–1003. DOI: 10.1172/JCI118884 49

Augst, A. D., Kong, H. J., Mooney, D. J. (2006). Alginate hydrogels as biomaterials. *Macromol Biosci* 6(8), 623–33. DOI: 10.1002/mabi.200600069 60

Baer, A. E., Laursen, T. A., Guilak, F., Setton, L. A. (2003). The micromechanical environment of intervertebral disc cells determined by a finite deformation, anisotropic, and biphasic finite element model. *J Biomech Eng* 125(1), 1–11. DOI: 10.1115/1.1532790 47

Baer, A. E., Wang, J. Y., Kraus, V. B., Setton, L. A. (2001). Collagen gene expression and mechanical properties of intervertebral disc cell–alginate cultures. *J Orthop Res* 19, 2–10. DOI: 10.1016/S0736-0266(00)00003-6 59, 60

Balakrishnan, B., Jayakrishnan, A. (2005). Self-cross-linking biopolymers as injectable in situ forming biodegradable scaffolds, *Biomaterials* 26, 3941–51. DOI: 10.1016/j.biomaterials.2004.10.005 65

Balazs, E. A. (1983). Sodium hyaluronate and viscosurgery. *In*: Healon (Sodium Hyaluronate): A Guide to Its Use in Ophthalmic Surgery (D. Miller and R. Stegmann, eds.), John Wiley & Sons, New York, pp. 5–28.

Bao, Q. B., Mccullen, G. M., Higham, P. A., Dumbleton, J. H., Yuan, H. A. (1996). The artificial disc: theory, design and materials. *Biomaterials* 17(12), 1157–67. DOI: 10.1016/0142-9612(96)84936-2 49

Baogan, P., Shuxun, H., Qi, S., Lianshun, J. (2001). The relationship between cartilage end-plate calcification and disc degeneration: an experimental study. *Chin Med J* 114, 308–12, 49

Bayliss, M., Johnstone, B., O'Brien, J. (1988). Proteoglycan synthesis in the human intervertebral disc. Variation with age, region and pathology. *Spine* 13(9), 972–81. DOI: 10.1097/00007632-198809000-00003 46

Beard, H. K., Roberts, S., O'Brien, J. P. (1981). Immunofluorescent staining for collagen and proteoglycan in normal and scoliotic intervertebral discs. *J Bone Joint Surg Am* 63B(4), 529–34. 46

Bertagnoli, R., Yue, J. J., Shah, R. V., Nanieva, R., Pfeiffer, F., Fenk-Mayer, A., Kershaw, T., Husted, D. S. (2005). The treatment of disabling single-level lumbar discogenic low back pain with total arthroplasty utilizing the Prodisc prosthesis. A prospective study with 2-year minimum follow-up. *Spine* 30(19), 2230–36, DOI: 10.1097/01.brs.0000182217.87660.40 50

Bertram, H., Kroeber, M., Wang, H., Unglaub, F., Guehring, T., Carstens, C., Richter, W. (2005). Matrix-assisted cell transfer for intervertebral disc cell therapy. *Biochem Biophys Res Commun* 331(4), 1185–92. DOI: 10.1016/j.bbrc.2005.04.034 67

Betre, H., Ong, S. R., Guilak, F., Chilkoti, A., Fermor, B., Setton, L.A. (2006). Chondrocytic differentiation of human adipose-derived adult stem cells in elastin-like polypeptide. *Biomaterials* 27(1), 91–9. DOI: 10.1016/j.biomaterials.2005.05.071

Betre, H., Setton, L. A., Meyer, D. E., Chilkoti, A. (2002). Characterization of a genetically engineered elastin-like polypeptide for cartilaginous tissue repair. *Biomacromolecules* 3(5), 910–6. DOI: 10.1021/bm0255037

Bibby, S. R., Jones, D. A., Lee, R. B., Yu, J., Urban, J. P. G. (2001). The pathophysiology of the intervertebral disc. Joint Bone Spine 68(6), 537–42. DOI: 10.1016/S1297-319X(01)00332-3 47, 49

Bibby, S., Jones, D., Ripley, R., Urban, J. (2005). Metabolism of the intervertebral disc. Effects of low levels of glucose, and pH on rates of energy metabolism of bovine nucleus pulposus cells. *Spine* 30, 487–96. DOI: 10.1097/01.brs.0000154619.38122.47 68

Biering, S. R. F. (1982). Low back trouble in a general population of 30-, 40-, 50-, and 60-year-old men and women: study design, representatives and basic results. *Dan Med Bull* 29(6), 289–99. 43

Blumenthal, S. L., Ohnmeiss, D. D., Guyer, R., Hochschuler, S., McAfee, P., Garcia, R., Salib, R., Yuan, H., Lee, C., Bertagnoli, R., Bryan, V., Winter, R. (2002). Artificial intervertebral discs and beyond: North American Spine Society Annual Meeting symposium. *Spine J* 2, 460–63. DOI: 10.1016/S1529-9430(02)00540-5 49

Blumenthal, S., McAfee, P.C., Guyer, R.D., Hochschuler, S.H., Geisler, F.H., Holt, R.T., Garcia, R. Jr, Regan, J.J., Ohnmeiss, D.D. (2005). A prospective, randomized, multicenter food and drug administration investigational device exemptions study of lumbar total disc replacement with the CHARITE artificial disc versus lumbar fusion. Part I. Evaluation of clinical outcomes. *Spine* 30(14), 1565–75. DOI: 10.1097/01.brs.0000170587.32676.0e 50

Boccaccini A.R., Blaker J.J. (2005). Bioactive composite materials for tissue engineering scaffolds. *Expert Rev Med Devices* 2(3), 303–17. DOI: 10.1586/17434440.2.3.303 66

Boelen, E. J. H., van Hooy-Corstjens, C. S. J., Bulstra, S. K., van Ooij, A., van Rhijn, L. W., Koole, L. H. (2005). Intrinsically radiopaque hydrogels for nucleus pulposus replacement. *Biomaterials* 26(33), 6674–83. DOI: 10.1016/j.biomaterials.2005.04.020

Boland, E. D., Matthews, J. A., Pawlowski, K. J., Simpson, D. G., Wnek, G. E., Bowlin, G. L. (2004). Electrospinning collagen and elastin: preliminary vascular tissue engineering, *Front Biosci* 9, 1422–32. DOI: 10.2741/1313

Boos, N., Semmer, N., Elfering, A., Schade, V., Gal, I., Zanetti, M., Kissling, R., Buchegger, N., Hodler, J., Main, C. J. (2000). Natural history of individuals with asymptomatic disc abnormalities in magnetic resonance imaging: predictors of low back pain-related medical consultation and work incapacity. *Spine* 25(12), 1484–92. DOI: 10.1097/00007632-200006150-00006 43

Boos, N., Weissbach, S., Rohrbach, H., Weiler, C., Spratt, K., Nerlich, A. (2002). Classification of age-related changes in lumbar intervertebral discs. *Spine* 27(23), 2631–44. DOI: 10.1097/00007632-200212010-00002 49

Borenstein, D. (1992). Epidemiology, etiology, diagnostic evaluation, and treatment of low back pain. *Curr Opin Rheumatol* 4(2), 226–32. DOI: 10.1097/00002281-199204000-00016 43

Bowles, R. D., Williams, R. M., Zipfel, W.R., Bonassar, L. J. (2010). Self-assembly of aligned tissue-engineered annulus fibrosus and intervertebral disc composite via collagen gel contraction. *Tissue Eng A* 16(4), 1339–48. DOI: 10.1089/ten.tea.2009.0442 60

Boyd, L. M., Carter, A. J. (2006). Injectable biomaterials and vertebral endplate treatment for repair and regeneration of the intervertebral disc. *Eur Spine J* 15(3), S414–21. DOI: 10.1007/s00586-006-0172-2 50

Brisby, H., Tao, H., Ma, D., Diwan, A. (2004). Cell therapy for disc degeneration-potentials and pitfalls. *Orthop Clin North Am* 35(1), 85–93. DOI: 10.1016/S0030-5898(03)00104-4 52

Broberg, K. B. (1983). On the mechanical behaviour of the intervertebral disc. *Spine* 8(2), 151–65. DOI: 10.1097/00007632-198303000-00006 46

Bron, J. L., Koenderink, G. H., Everts, V., Smit, T. H. (2009). Rheological characterization of the nucleus pulposus and dense collagen scaffolds intended for functional replacement. *J Orthop Res* 27(5): 620–626. DOI: 10.1002/jor.20789

Brown, R. Q., Mount, A., Burg K. J. (2005). Evaluation of polymer scaffolds to be used in a composite injectable system for intervertebral disc tissue engineering. *J Biomed Mater Res A* 74(1), 32–9. DOI: 10.1002/jbm.a.30250 65

Brown, R. Q., Mount, A., Burg, K. J. L. (2004). Evaluation of polymer scaffolds to be used in a composite injectable system for intervertebral disc tissue engineering. *J Biomed Mater Res* 74A, 32–36, DOI: 10.1002/jbm.a.30250

Bruehlmann, S. B., Rattner, J. B., Matyas, J. R., Duncan, N. A. (2002). Regional variations in the cellular matrix of the annulus fibrosus of the intervertebral disc. *J Anat* 201, 159–165, DOI: 10.1046/j.1469-7580.2002.00080.x 47

Bryant, S. J., Durand, K. L., Anseth, K. S. (2003). Manipulations in hydrogel chemistry control photoencapsulated chondrocyte behavior and their extracellular matrix production. *J Biomed Mater Res A* 67:1430–6. DOI: 10.1002/jbm.a.20003 65

Buckwalter, J. (1995). Spine update. Aging and degeneration of the human intervertebral disc. *Spine* 20, 1307–13, 46, 49

Buckwalter, J. A., Pedrini-Mille, A., Pedrini, V., Tudisco, C. (1985). Proteoglycans of human infant intervertebral disc. Electron microscopic and biochemical studies. *J Bone Joint Surg Am* 67(2), 284–94. 46

Buckwalter, J.A., Smith, K.C., Kazarien, L.E., Rosenberg, L.C., Ungar, R. (1989). Articular cartilage and intervertebral disc proteoglycans differ in structure. *J Orthop Res* 7, 146–51. DOI: 10.1002/jor.1100070121 46

Burdick, J. A., Chung, C., Jia, X., Randolph, M. A., Langer, R. (2005). Controlled degradation and mechanical behavior of photopolymerized hyaluronic acid networks. *Biomacromolecules* 6, 386–91. DOI: 10.1021/bm049508a 65

Burkus, J. (2002). Advances in spine surgery. Intervertebral fixation: clinical results with anterior cages. *Orthop Clin North Am* 33(2), 349–57. DOI: 10.1016/S0030-5898(01)00012-8 49

Burkus, J., Schuler, T., Gornet, M., Zdeblick, T. (2004). Anterior lumbar interbody fusion for the management of chronic lower back pain: current strategies and concepts. *Orthop Clin North Am* 35(1), 25–32. DOI: 10.1016/S0030-5898(03)00053-1 49

Cassinelli, E. H., Hall, R. A., Kang, J. D. (2001). Biochemistry of intervertebral disc degeneration and the potential for gene therapy applications. *Spine J* 1(3), 205–14. DOI: 10.1016/S1529-9430(01)00021-3 49

Bushell, G. R., Ghosh, P., Taylor, T. F. K., Akeson, W. H. (1977). Proteoglycan chemistry of the intervertebral disc. *Clin Orthop Rel Res* 129, 115–24. 46

Chang, S., Masuda, K., Takegami, K., Sumner, D., Thonar, E. J. M. A., Andersson, G., An, H. (2000). Gene gun–mediated gene transfer to intervertebral disc cells. *Trans. Orthop. Res. Soc.* 25, 231–37.

Cheng, Y. H., Lin, F. H., Yang, K. C. (2009). Thermosensitive chitosan-gelatin-glycerol phosphate hydrogels as a cell carrier for nucleus pulposus regeneration: an *in-vitro* study. *Tissue Eng Part A* (Epub)

Chiba, K., Andersson, G. B., Masuda, K., Thonar, E. J. (1997). Metabolism of the extra-cellular matrix formed by intervertebral disc cells cultured in alginate. *Spine* 22, 2885–92. DOI: 10.1097/00007632-199712150-00011 59

Chou, A. I., Akintoye, S. O., Nicoll, S. B. (2009). Photo-crosslinked alginate hydrogels support enhanced matrix accumulation by nucleus pulposus cells *in vivo*. *Osteoarthritis Cartilage* 17(10), 1377–84. DOI: 10.1016/j.joca.2009.04.012

Chou, A. I., Nicoll, S. B. (2009). Characterization of photocrosslinked alginate hydrogels for nucleus pulposus cell encapsulation. *J Biomed Mater Res A* 91(1), 187–94. DOI: 10.1002/jbm.a.32191 59

Cloyd, J. M., Malhotra, N. R., Weng, L., Chen, W., Mauck, R. L., Elliott, D. M. (2007). Material properties in unconfined compression of human nucleus pulposus, injectable hyaluronic acid-based hydrogels and tissue engineering scaffolds. Eur *Spine J* 16(11), 1892–98. DOI: 10.1007/s00586-007-0443-6 65, 66, 67

Coventry, M. B., Ghormley, R. K., Kernohan, J. W. (1945a). The intervertebral disc: Its microscopic anatomy and pathology, Part I: anatomy, development, and physiology. *J. Bone Joint Surg Am* 27, 105–12. 45

Crean, J., Roberts, S., Jaffray, D., Eisenstein, S., Duance, V. (1997). Matrix metalloproteinases in the human intervertebral disc: role in disc degeneration and scoliosis. *Spine* 22, 2877–83. DOI: 10.1097/00007632-199712150-00010 49

Crevensten, G., Walsh, A .J., Ananthakrishnan, D., Page, P., Wahba, G. M., Lotz, J. C., Berven, S. (2004). Intervertebral disc cell therapy for regeneration: mesenchymal stem cell implantation in rat intervertebral discs. *Ann Biomed Eng* 32(3), 430–34. DOI: 10.1023/B:ABME.0000017545.84833.7c 65, 67

Dang, J. M., Leong, K. W. (2006). Natural polymers for gene delivery and tissue engineering. *Adv Drug Deliv Rev.* 58(4), 487–99. DOI: 10.1016/j.addr.2006.03.001

Darwis, D., Stasica, P., Razzak, M. T., Rosiak, J. M. (2002). Characterization of poly (vinyl alcohol) hydrogel for prosthetic intervertebral disc nucleus. *Rad Phys Chem* 63, 539–46. DOI: 10.1016/S0969-806X(01)00636-3 49

Debelle, L., Alix, A. J., Wei, S. M., Jacob, M. P., Huvenne, J. P., Berjot, M., Legrand, P. (1998). The secondary structure and architecture of human elastin. *Eur J Biochem* 258, 533–9. DOI: 10.1046/j.1432-1327.1998.2580533.x

Desai, B. J., Gruber, H. E., Hanley, E. N. Jr. (1999). The influence of matrigel or growth factor reduced matrigel on human intervertebral disc cell growth and proliferation. *Histol Histopathol* 14, 359–407.

Deyo, R. A., Tsui-Wu, Y. J. (1987). Descriptive epidemiology of low back pain and its related medical care in the United States. *Spine* 12, 264–71. 43

Di Martino, A., Sittinger, M., Risbud, M. V. (2005b). Chitosan: A versatile biopolymer for orthopaedic tissue-engineering. *Biomaterials* 26(30), 5983–90. DOI: 10.1016/j.biomaterials.2005.03.016 68

Di Martino, A., Vaccaro, A. R., Lee, J. Y., Denaro, V., Lim, M. R. (2005a). Nucleus pulposus replacement. Basic science and indications for clinical use. *Spine* 30, S16. DOI: 10.1097/01.brs.0000174530.88585.32 50

Driesse, M. J., Esandi, M. C., Kros, J. M., Avezaat, C. J., Vecht, C., Zurcher, C., van der Velde, I., Valerio, D., Bout, A., and Sillevis Smitt, P. A. (2000). Intra-CSF administered recombinant adenovirus causes an immune response–mediated toxicity. *Gene Ther* 7(16), 1401–09. DOI: 10.1038/sj.gt.3301250

East, G. C., McIntyre, J. E., Qin, Y. (1989). Medical use of chitosan. *In* "Chitin and Chitosan" (Skjak-Braek, G., Anthonsen, T., Sandford, P. eds.), Elsevier, London, pp. 757–764.

Ebara, S., Iatridis, J. C., Setton, L. A., Foster, R. J., Mow, V. C., Weidenbaum, M. (1996). Tensile properties of nondegenerate human lumbar anulus fibrosus. *Spine* 21(4), 452–61. DOI: 10.1097/00007632-199602150-00009 44

Ehrlich, G. E. (2003). Low back pain. *Bull World Health Org* 81(9), 671–76. 43

Elfering, A., Semmer, N., Birkhofer, D., Zanetti, M., Hodler, J., Boos, N. (2002). Risk factors for lumbar disc degeneration: a 5-year prospective MRI study in asymptomatic individuals. *Spine* 27, 125–34. DOI: 10.1097/00007632-200201150-00002 43

Elisseeff, J. H., Anseth, K., Sims, D., McIntosh, W., Randolph, M., Langer, R. (1999). Transdermal photopolymerization for minimally invasive implantation. *Proc Natl Acad Sci USA* 16, 3104–7. DOI: 10.1073/pnas.96.6.3104 65

Errington, R. J., Puustjarvi, K., White, I. R., Roberts, S., Urban, J. P. (1998). Characterisation of cytoplasm-filled processes in cells of the intervertebral disc. *J Anat* 192, 369–78. DOI: 10.1046/j.1469-7580.1998.19230369.x 47

Erwin, W. M., Las Heras, F., Islam, D., Fehlings, M. G., Inman, R. D. (2009). The regenerative capacity of the notochordal cell: tissue constructs generated in vitro under hypoxic conditions. *J Neurosurg Spine* 10(6), 513–21. DOI: 10.3171/2009.2.SPINE08578 60

Eyre, D. R. (1979). Biochemistry of the intervertebral disc. *Inter Rev Conn Tis Res* 8, 227–91. 46

Eyre, D. R. (1988). Collagens of the disc. In: Ghosh, P., ed. The Biology of the Intervertebral Disc, Vol. 1. Boca Raton, FL: CRC Press, pp. 171–188. 46

Eysel, P., Rompe, J., Schoenmayr, R., Zoellner, J. (1999). Biomechanical behaviour of a prosthetic lumbar nucleus. *Acta Neurochir* 141(10), 1083–7. DOI: 10.1007/s007010050486 50

Fraylich, M. R., Liu, R., Richardson, S. M., Baird, P., Hoyland, J., Freemont, A. J., Alexander, C., Shakesheff, K., Cellesi, F., Saunders, B. R. (2010). Thermally-triggered gelation of PLGA dispersions: towards an injectable colloidal cell delivery system. *J Colloid Interface Sci* 344(1), 61–9. DOI: 10.1016/j.jcis.2009.12.030 59

Frymoyer, J. W., Cats-Baril, W. L. (1991). An overview of the incidences and costs of low back pain. *Orthop Clin North Am* 22(2), 263–71. 43

Gaetani, P., Torre, M. L., Klinger, M., Faustini, M., Crovato, F., Bucco, M., Marazzi, M., Chlapanidas, T., Levi, D., Tancioni, F., Vigo, D., Rodriguez y Baena, R. (2008). Adipose-derived stem cell therapy for intervertebral disc regeneration: an *in vitro* reconstructed tissue in alginate capsules. *Tissue Eng Part A* 14(8), 1415–23. DOI: 10.1089/ten.tea.2007.0330 60

Goupille, P., Jayson, M., Valat, J., Freemont, A. (1998). Matrix metalloproteinases: the clue to intervertebral disc degeneration? *Spine* 23(14), 1612–26. 49

Ghosh, P., Bushell, G. R., Taylor, R. K. F., Pearce, R. H., Grimmer, B. J. (1977). Distribution of glycosaminoglycan across the normal and scoliotic disc. *Spine* 5, 310–26. 46

Grange, L., Gaudin, P., Trocme, C., Phelip, X., Morel, F., Juvin, R. (2001). Intervertebral disk degeneration and herniation. The role of metalloproteinases and cytokines. Joint Bone *Spine* 68(6), 547–53. DOI: 10.1016/S1297-319X(01)00324-4 49

Gruber, H. E., Fisher, E. C., Jr., Desai, B., Stasky, A. A., Hoelscher, G., and Hanley, E. N., Jr. (1997a). Human intervertebral disc cells from the anulus: three-dimensional culture in agarose or alginate and responsiveness to TGF-beta1. *Exp Cell Res* 235(1), 13–21. DOI: 10.1006/excr.1997.3647 52

Gruber, H. E., Leslie, K., Ingram, J., Norton, H.J., Hanley, J., Edward, N., (2004). Cell-based tissue engineering for the intervertebral disc: *in vitro* studies of human disc cell gene expression and matrix production within selected cell carriers. *Spine J* 4(1), 44–55. DOI: 10.1016/S1529-9430(03)00425-X 54, 60

Gruber, H. E., Stasky, A. A., Hanley, E. N. Jr. (1997b). Characterization and phenotypic stability of human disc cells *in vitro*. *Matrix Biol* 16(5), 285–8. DOI: 10.1016/S0945-053X(97)90016-0 59

Gruber, H. E., Hanley, E. N. Jr. (2003b). Recent advances in disc cell biology. *Spine* 28(2), 186–93. DOI: 10.1097/00007632-200301150-00017 47, 54

Gruber, H. E., Hanley, E. N. Jr. (2003a). Biologic strategies for the therapy of intervertebral disc degeneration. *Expert Opin Biol Ther* 3, 1209–14. DOI: 10.1517/14712598.3.8.1209 47

Gruber, H.E., Johnson, T.L., Leslie, K., Ingram, J.A., Martin, D., Hoelscher, G., Banks, D., Phieffer, L., Coldham, G., Hanley, E.N. Jr. (2002). Autologous intervertebral disc cell implantation: a model using psammomys obesus, the sand rat. *Spine* 27(15), 1626–33. DOI: 10.1097/00007632-200208010-00007

Gu, Z., Ma, Y., Gao, J., Liu, J., Li, Y. (2004). Development of artificial disc nucleus materials (semicrystalline polyvinyl alcohol hydrogel elastomers). *Sheng Wu Yi Xue Gong Cheng Xue Za Zhi* 21(3): 347–49.

Guiot, B.H., Fessler, R.G. (2000). Molecular biology of degenerative disc disease. *Neurosurgery* 47(5), 1034–40. DOI: 10.1097/00006123-200011000-00003 43

Gunatillake, P. A., Adhikari, R. (2003). Biodegradable synthetic polymers for tissue engineering. *Eur Cell Mater* 5, 1–16. 59

Ha, H. J., Kim, S. H., Yoon, S. J., Park, S. W., So, J. W., Rhee, J. M., Kim, M. S., Khang, G., Lee, H. B. (2008). Evaluation of various scaffolds for tissue engineered biodisc using annulus fibrosus cells. *Polymer-Korea* 32(1), 26–30.

Haberstroh, K., Enz, A., Zenclussen, M. L., Hegewald, A. A., Neumann, K., Abbushi, A., Thome, C., Sittinger, M., Endres, M., Kaps, C. (2009). Human intervertebral disc-derived cells are recruited by human serum and form nucleus pulposus-like tissue upon stimulation with TGF-beta 3 or hyaluronan *in vitro*. *Tissue & Cell* 41(6), 414–20. DOI: 10.1016/j.tice.2009.05.006 65

Halloran, D.O., Grad, S., Stoddart, M., Dockery, P., Alini, M., Pandit, A.S. (2008). An injectable cross-linked scaffold for nucleus pulposus regeneration. *Biomaterials* 29(4), 438–47. DOI: 10.1016/j.biomaterials.2007.10.009 66

Hayes, A. J., Benjamin, M., Ralphs, J. R. (2001). Extra cellular matrix development of the intervertebal disc. *Matrix Biol* 20, 107–23. DOI: 10.1016/S0945-053X(01)00125-1 46

Healthcare Cost and Utilization Project, (2003) National Statistics [on-line]. Available at www.hcup.ahrq.gov Accessed on July 17, 2005. 49

Heinegard, D., Paulsson, M. (1980). Proteoglycans and matrix proteins in cartilage. *In:* The Biochemistry of Glycoproteins and Proteoglycans (Lennarz, W. J. ed.), Plenum Press, New York, pp. 297–328. 65

Henriksson, H. B., Svanvik, T., Jonsson, M., Hagman, M., Horn, M., Lindahl, A., Brisby, H. (2009). Transplantation of human mesenchymal stems cells into intervertebral discs in a xenogeneic porcine model. *Spine* 34(2), 141–8. DOI: 10.1097/BRS.0b013e31818f8c20 67

Hickey, D. (1980). X-ray diffraction studies of the arrangement of collagenous fibres in human fetal intervertebral disc. *J Anat* 131, 81–90. 44

Holladay, C., Keeney, M., Greiser, U., Murphy, M., O'Brien, T., Pandit, A. (2009). Matrix reservoir for improved control of non-viral gene delivery. *J. Control Release* 136(3), 220–5. DOI: 10.1016/j.jconrel.2009.02.006

Hong, H. K., Kim, S. F., Lee, S. K., Lee, Y. H., Kim, S. J., Kim, O. Y., Lee, D., Rhee, J. M., Son, Y., Khang, G. (2009). Compressive strength of poly(L-lactide-co-glycolide) scaffolds seeded nucleus pulposus cells depending on pore size. *Tissue Engineering and Regenerative Medicine* 6(4–11), 1029-34. 59

Horner, H., Urban, J. (2001). Effect of nutrient supply on the viability of cells from the nucleus pulposus of the intervertebral disc. *Spine* 26(23), 2543–9. DOI: 10.1097/00007632-200112010-00006 49

Huang, B., Li, C. Q., Zhou, Y., Luo, G., Zhang, C. Z. (2010). Collagen II/hyaluronan/chondroitin-6-sulfate tri-copolymer scaffold for nucleus pulposus tissue engineering. *J Biomed Mater Res B Appl Biomater* 92(2), 322–31. DOI: 10.1002/jbm.b.31518 66

Huang, R., Sandhu, H. (2004). The current status of lumbar total disc replacement. *Orthop Clin North Am* 35(1), 33–42. DOI: 10.1016/S0030-5898(03)00103-2 49

Hukins, D. W. L. (1988). Disc structure and function. *In*: The biology of the intervertebral disc. (Ghosh, P., ed.), Vol. 1. Boca Raton, FL: CRC Press, pp. 1–38. 45, 47

Humzah, M., Soames, R. (1988). Human intervertebral disc: structure and function. *Anat Rec* 220, 337–56. DOI: 10.1002/ar.1092200402 44

Hunter, C., Matyas, J., Duncan, N. (2003). The notochordal cell in the nucleus pulposus: a review in the context of tissue engineering. *Tissue Eng* 9, 667–77. DOI: 10.1089/107632703768247368 52

Hutton, W.C., Toribatake, Y., Elmer, W.A., Ganey, T.M., Tomita, K., Whitesides, T.E. The effect of compressive force applied to the intervertebral disc *in vivo*. (1998). A study of proteoglycans and collagen. *Spine* 23, 2524–37. 43

Ikada, Y., Tabata, Y. (1998). Protein release from gelatin matrices. *Adv. Drug Del Rev.* 31, 287–301. DOI: 10.1016/S0169-409X(97)00125-7 65

Inerot, S., Axelsson, I. (1991). Structure and composition of proteoglycans from human annulus fibrosus. *Connect Tissue Res* 26 (1–2), 47-63. DOI: 10.3109/03008209109152163 44

Jung, S. H., Jang, J. W., Kim, S. H., Hong, H. H., Oh, A. Y., Rhee, J. M., Kang, Y. S., Khang, G. (2008). Articular cartilage regeneration using hyaluronic acid loaded PLGA scaffold by emulsion freeze-drying method. *Tissue Engineering and Regenerative Medicine* 5(4–6), 643-49. 65

Jünger, S., Gantenbein-Ritter, B., Lezuo, P., Alini, M., Ferguson, S. J., Ito, K. (2009). Effect of limited nutrition on in situ intervertebral disc cells under simulated-physiological loading. *Spine* 34(12), 1264–71. DOI: 10.1097/BRS.0b013e3181a0193d 67

Kabanov, A. V., Kabanov, V. A. (1995). DNA complexes with polycations for the delivery of genetic material into cells. *Bioconjug Chem* 6(1), 7–20. DOI: 10.1021/bc00031a002

Kandel, R., Roberts, S., Urban, J. P. (2008). Tissue engineering and the intervertebral disc: the challenges. *Eur Spine J* 17(4), 480–91. DOI: 10.1007/s00586-008-0746-2 49

Kanemoto, M., Hukuda, S., Komiya, Y., Katsuura, A., Nishioka, J. (1996). Immunohistochemical study of matrix metalloproteinase-3 and tissue inhibitor of metalloproteinase-1 in human intervertebral discs. *Spine* 21(1), 1–8. DOI: 10.1097/00007632-199601010-00001 49

Kaplan, D. L., Wiley, B. J., Mayer, J. M., Arcidiacono, S., Keith, J., Lombardi, S. J., Ball, D. and Allen, A. L. (1994). Biosynthetic polysaccharides. *In:* Biomedical Polymers (Shalaby, S. ed.), Hanser, New York, pp. 189–212.

Kawaguchi, Y., Osada, R., Kanamori, M., Ishihara, H., Ohmori, K., Matsui, H., Kimura, T. (1999). Association between an aggrecan gene polymorphism and lumbar disc degeneration. *Spine* 24, 2456–61. DOI: 10.1097/00007632-199912010-00006 49

Kawai, K., Suzuki, S., Tabata, Y., Ikada, Y., Nishimura, Y. (2000). Accelerated tissue regeneration through incorporation of basic fibroblast growth factor-impregnated gelatin microspheres into artificial dermis, *Biomaterials* 21, 489–99. DOI: 10.1016/S0142-9612(99)00207-0 65

Keeney, M., van den Beucken, J. J., van der Kraan, P. M., Jansen, J. A., Pandit, A. (2010). The ability of a collagen/calcium phosphate scaffold to act as its own vector for gene delivery and to promote bone formation via transfection with VEGF165. *Biomaterials* 31(10), 2893–902. DOI: 10.1016/j.biomaterials.2009.12.041

Kelsey, J.L., White, A.A. (1980). Epidemiology and impact of low-back pain. *Spine* 5(2), 133–42. DOI: 10.1097/00007632-198003000-00007 43

Kim, D. J., Moon, S. H., Kim, H., Kwon, U. H., Park, M. S., Han, K. J., Hahn, S. B., and Lee, H. M. (2003). Bone morphogenetic protein-2 facilitates expression of chondrogenic, not osteogenic, phenotype of human intervertebral disc cells. *Spine* 28(24), 2679–84. DOI: 10.1097/01.BRS.0000101445.46487.16

Kim, H., Lee, J. U., Moon, S. H., Kim, H. C., Kwon, U. H., Seol, N. H., Kim, H. J., Park, J. O., Chun, H. J., Kwon, I. K. (2009). Zonal responsiveness of the human intervertebral disc to bone morphogenetic protein-2. *Spine* 34(17), 1834–38. DOI: 10.1097/BRS.0b013e3181ae18ba 59, 60

Kim, T. H., Jiang, H. L., Nah, J. W., Cho, M. H., Akaike, T., Cho, C. S. (2007). Receptor-mediated gene delivery using chemically modified chitosan. *Biomed Mater* 2(3), S95–100. DOI: 10.1088/1748-6041/2/3/S02

Kitahara, H. (1979). Histochemical study of the human intervertebral disc. *Nippon Seikeigeka Gakkai Zasshi* 53(7), 817–30. 46

Klara, P., Ray, C. (2002). Artificial nucleus replacement: clinical experience. *Spine* 27(12), 1374–7, DOI: 10.1097/00007632-200206150-00022 50

Kobayashi, Y., Sakai, D., Iwashina, T., Iwabuchi, S., Mochida, J. (2009). Low-intensity pulsed ultrasound stimulates cell proliferation, proteoglycan synthesis and expression of growth factor-related genes in human nucleus pulposus cell line. *Eur Cell Mater* 17, 15–22. 60

Korecki, C. L., Kuo, C. K., Tuan, R. S., Iatridis, J. C. (2009). Intervertebral disc cell response to dynamic compression is age and frequency dependent. *J Orthop Res* 27(6), 800–6. DOI: 10.1002/jor.20814 60

Kuijpers, A. J., van Wachem, P. B., van Luyn, M. J., Plantinga, J. A., Engbers, G. H., Krijgsveld, J., Zaat, S. A., Dankert J. Feijen, J. (2000a). *In vivo* compatibility and degradation of crosslinked gelatin gels incorporated in knitted Dacron. *J Biomed Mater Res* 51, 136–45. DOI: 10.1002/(SICI)1097-4636(200007)51:1%3C136::AID-JBM18%3E3.0.CO;2-W 65

Kuijpers, A.J., Engbers, G.H., Krijgsveld, J., Zaat, S.A., Dankert, J., Feijen, J. (2000b). Cross-linking and characterisation of gelatin matrices for biomedical applications, *J Biomater Sci, Polym Ed* 11, 225–43. DOI: 10.1163/156856200743670 65

Kulkarni, M., Greiser, U., O'Brien, T., Pandit, A. (2010). Liposomal gene delivery mediated by tissue-engineered scaffolds. *Trends Biotechnol* 28(1), 28–36. DOI: 10.1016/j.tibtech.2009.10.003

Kuslich, S. D., Ulstrom, C. L., Michael, C. J. (1991). The tissue origin of low back pain and sciatica: a report of pain response to tissue stimulation during operations on the lumbar spine using local anesthesia. *Orthop Clin North Am* 22, 181–7. 43

Kwon, B., Vaccaro, A., Grauer, J., Beiner, J. (2003). Indications, techniques, and outcomes of posterior surgery for chronic low back pain. *Orthop Clin North Am* 34(2), 297–308. DOI: 10.1016/S0030-5898(03)00014-2 49

Laleg, M., Pikulik, I. (1991). Wet-web strength increase by chitosan. *Nordic Pulp Paper Res J* 9, 99–103. DOI: 10.3183/NPPRJ-1991-06-03-p099-103

Lamme, E. N., van Leeuwen, R. T., Jonker, A., van Marle, J., Middelkoop, E. (1998). Living skin substitutes: survival and function of fibroblasts seeded in a dermal substitute in experimental wounds. *J Invest Dermatol* 111, 989–95. DOI: 10.1046/j.1523-1747.1998.00459.x

Langer, R., Tirrell D. A. (2004). Designing materials for biology and medicine. *Nature* 428(6982), 487–92. DOI: 10.1038/nature02388

Langrana, N., Parsons, J., Lee, C., Vuono-Hawkins, M., Yang, S., Alexander, H. (1994). Materials and design concepts for an intervertebral disc spacer. I. Fiber-reinforced composite design. *J Appl Biomat* 5(2), 125–32. DOI: 10.1002/jab.770050205 49

Lanir, Y., Seybold, J., Schneiderman, R., Huyghe, J. M. (1998). Partition and diffusion of sodium and chloride ions in soft charged foam: the effect of external salt concentration and mechanical deformation. *Tissue Eng* 4(4): 365–78. DOI: 10.1089/ten.1998.4.365

Le Maitre, C. L., Frain, J., Fotheringham, A. P., Freemont, A. J., Hoyland, J. A. (2008). Human cells derived from degenerate intervertebral discs respond differently to those derived from non-degenerate intervertebral discs following application of dynamic hydrostatic pressure. *Biorheology* 45(5), 563–75. DOI: 10.3233/BIR-2008-0498 60

Le Maitre, C. L., Freemont, A. J., and Hoyland, J. A. (2006). A preliminary *in vitro* study into the use of IL-1Ra gene therapy for the inhibition of intervertebral disc degeneration. *Int J Exp Pathol* 87(1), 17–28. DOI: 10.1111/j.0959-9673.2006.00449.x 68

Le Visage, C., Yang, S. H., Kadakia, L., Sieber, A. N., Kostuik, J. P., Leong, K. W. (2004). Small intestinal submucosa (SIS) as a potential bioscaffold for interverte-bral disc regeneration. Abstract Presented NASS 19th Annual Meeting, *Spine J* 4, 3S. DOI: 10.1097/01.brs.0000238684.04792.eb

Lee, C. R., Grad, S., Gorna, K., Gogolewski, S., Goessl, A., Alini, M. (2005). Fibrin-polyurethane composites for articular cartilage tissue engineering: a preliminary analysis. *Tissue Eng* 11(9–10), 1562-73. DOI: 10.1089/ten.2005.11.1562 59

Lee, H., Sowa, G., Vo, N., Vadala, G., O'Connell, S., Studer, R., Kang, J. (2010). Effect of bupivacaine on intervertebral disc cell viability. *Spine J* 10(2), 159–66. DOI: 10.1016/j.spinee.2009.08.445 60

Lemaire, J. (1997). Intervertebral disc prosthesis. Results and prospects for the year 2000. *Clin Orthop Rel Res* 337, 64–76. 50

Leone, G., Torricelli, P., Chiumiento, A., Facchini, A., Barbucci, R. (2008). Amidic arginate hydrogel for nucleus pulposus replacement. *J Biomed Mater Res A* 84(2), 391–401. DOI: 10.1002/jbm.a.31334 67

Li, X., Leo, B. M., Beck, G., Balian, G., Anderson, G. D. (2004). Collagen and proteoglycan abnormalities in the GDF-5-deficient mice and molecular changes when treating disk cells with recombinant growth factor. *Spine* 29(20), 2229–34. DOI: 10.1097/01.brs.0000142427.82605.fb

Lim, D. W., Nettles, D. L., Setton, L. A., Chilkoti, A. (2008). *In situ* cross-linking of elastin-like polypeptide block copolymers for tissue repair. *Biomacromolecules* 9(1), 222–30. DOI: 10.1021/bm7007982

Lipson, S. J., Muir, H. (1981). Experimental intervertebral disc degeneration: morphologic and proteoglycan changes over time. *Arthritis Rheum* 24(1), 12–21. DOI: 10.1002/art.1780240103 43

Liu, J., Wang, J., Zhou, Y. (2009). BMSCs-chitosan hydrogel complex transplantation for treating intervertebral disc degeneration. *Zhongguo Xiu Fu Chong Jian Wai Ke Za Zhi* 23(2), 178–82. 67

Lotz, J. C., Chin, J. R. (2000). Intervertebral disc cell death is dependent on the magnitude and duration of spinal loading. *Spine* 25, 1477–85. DOI: 10.1097/00007632-200006150-00005 43

Lu, Z. F., Doulabi, B. Z., Wuisman, P. I., Bank, R. A., Helder, M. N. (2008). Influence of collagen type II and nucleus pulposus cells on aggregation and differentiation of adipose tissue-derived stem cells. *J Cell Mol Med* 12(6B), 2812–22. DOI: 10.1111/j.1582-4934.2008.00278.x

Lyons, G., Eisenstein, S. M., Sweet, M. B. (1981). Biochemical changes in intervertebral disc degeneration. *Biochim Biophys Acta* 673(4), 443–53. DOI: 10.1016/0304-4165(81)90476-1 43

MacGregor, A. J., Andrew, T., Sambrook, P. N., Spector, T. D. (2004). Structural, psychological, and genetic influences on low back and neck pain: a study of adult female twins. *Arthritis Rheum* 51(2), 160–7. DOI: 10.1002/art.20236 49

MacLean, J. J., Lee, C. R., Grad, S., Ito, K., Alini, M., Iatridis, J. C. (2003). Effects of immobilization and dynamic compression on intervertebral disc cell gene expression *in vivo*. *Spine* 28(10), 973–81.

Maldonado B. A., Oegema T. R. Jr. (1992). Initial characterization of the metabolism of intervertebral disc cells encapsulated in microspheres. *J Orthop Res* 10, 677–90. DOI: 10.1002/jor.1100100510 47, 59

Masuda, K., An, H. S. (2004). Growth factors and the intervertebral disc. *Spine J* 4 (6 Suppl.), 330S-340S. DOI: 10.1016/j.spinee.2004.07.028

Masuda, K., Takegami, K., An, H., Kumano, F., Chiba, K., Andersson, G.B., Schmid, T., Thonar, E. (2003). Recombinant osteogenic protein-1 upregulates extracellular matrix metabolism by rabbit annulus fibrosus and nucleus pulposus cells cultured in alginate beads. *J Orthop Res* 21(5), 922–30. DOI: 10.1016/S0736-0266(03)00037-8

Mauth, C., Bono, E., Haas, S., Paesold, G., Wiese, H., Maier, G., Boos, N., Graf-Hausner, U. (2009). Cell-seeded polyurethane-fibrin structures-a possible system for intervertebral disc regeneration. *Eur Cell Mater* 18(7–38), 29-38. 54

McAfee, P.C., Cunningham, B., Holsapple, G., Adams, K., Blumenthal, S., Guyer, R.D., Dmietriev, A., Maxwell, J.H., Regan, J.J., Isaza, J. (2005). A prospective, randomized, multicenter food and drug administration investigational device exemptions study of lumbar total disc replacement with the CHARITE artificial disc versus lumbar fusion. Part II. Evaluation of radiographic outcomes and correlation of surgical technique accuracy with clinical outcomes. *Spine* 30(14), 1576–83. DOI: 10.1097/01.brs.0000170561.25636.1c 50

McHale, M. K., Setton, L. A., Chilkoti, A. (2005). Synthesis and *in vitro* evaluation of enzymatically cross-linked elastin-like polypeptide gels for cartilaginous tissue repair. *Tissue Eng* 11(11–12), 1768-79 DOI: 10.1089/ten.2005.11.1768

Melrose, J., Smith, S., Ghosh, P., Taylor, T. K. F. (2001). Differential expression of proteoglycan epitopes and growth characteristics of intervertebral disc cells grown in alginate bead culture. *Cells Tissues Organs* 168, 137–46. DOI: 10.1159/000047829 59

Mikos, A., Huygens G. (2003). *Lecture: Tissue Engineering,* Netherlands Organisation for Scientific Research.

Miller, J. A., Schmatz, C., Schultz, A. B. (1988). Lumbar disc degeneration: correlation with age, sex, and spine level in 600 autopsy specimens. *Spine* 13(2), 173–8. 43

Mithieux, S. M., Weiss, A. S. (2005). Elastin. *Adv Protein Chem* 70, 437–61. DOI: 10.1016/S0065-3233(05)70013-9

Mizuno, H., Roy, A. K., Vacanti, C. A., Kojima, K., Ueda, M., Bonassar, L. J. (2004). Tissue-engineered composites of annulus fibrosus and nucleus pulposus for intervertebral disc replacement. *Spine* 29(12), 1290–7. DOI: 10.1097/01.BRS.0000128264.46510.27 54

Moore, K., Pinto, M., Butler, L. (2002). Degenerative disc disease treated with combined anterior and posterior arthrodesis and posterior instrumentation. *Spine* 27(15), 1680–6. DOI: 10.1097/00007632-200208010-00018 49

Moore, R. (2000). The vertebral end-plate: what do we know? *Eur Spine J* 9(2), 92–96. DOI: 10.1007/s005860050217 45, 49

Mortisen, D., Peroglio, M., Alini, M., Eglin, D. (2010). Tailoring thermoreversible hyaluronan hydrogels by click chemistry and RAFT polymerization for cell and drug therapy. *Biomacromolecules* 11(5), 1261–72. DOI: 10.1021/bm100046n 65

Muzzarelli, R., Baldassara, V., Conti, F., Ferrara, P., Biagini, G., Gazzarelli, G., Vasi, V. (1988). Biological activity of chitosan: ultrastructural study. *Biomaterials* 9, 247–52. DOI: 10.1016/0142-9612(88)90092-0

Mwale, F., Iordanova, M., Demers, C. N., Steffen, T., Roughley, P., Antoniou, J. (2005). Biological evaluation of chitosan salts cross-linked to genipin as a cell scaffold for disk tissue engineering. *Tissue Eng* 11, 130–40. DOI: 10.1089/ten.2005.11.130

Mwale, F., Petit, A., Tian Wang, H., Epure, L. M., Girard-Lauriault, P. L., Ouellet, J. A., Wertheimer, M. R., Antoniou, J. (2008). The potential of N-rich plasma-polymerized ethylene (PPE:N) films for regulating the phenotype of the nucleus pulposus. *Open Orthop J* 2, 137–44. DOI: 10.2174/1874325000802010137

Nachemson, A., Lewin, T., Maroudas, A., Freeman, M. (1970). *In vitro* diffusion of dye through the endplates and the annulus fibrosus of human lumbar inter-vertebral discs. *Acta Orthop Scand* 41(6), 589–607. DOI: 10.3109/17453677008991550 45

Naeme, P. J., Barry, F. P. (1993). The link proteans. *Experimentia* 49, 393–402. 65

Nagae, M., Ikeda, T., Mikami, Y., Hase, H., Ozawa, H., Matsuda, K., Sakamoto, H., Tabata, Y., Kawata, M., Kubo, T. (2007). Intervertebral disc regeneration using platelet-rich plasma and biodegradable gelatin hydrogel microspheres. *Tissue Eng* 13(1), 147–58. DOI: 10.1089/ten.2006.0042

Natarajan, R. N., Ke, J. H., Andersson, G. B. (1994). A model to study the disc degeneration process. *Spine* 19(3), 259–65. DOI: 10.1097/00007632-199402000-00001 43

Neame, P. J., Kay, C. J., McQuillan, D. J., Beales, M. P., Hassell, J. R. (2000). Independent modulation of collagen fibrillogenesis by decorin and lumican. *Cell Mol Life Sci* 57(5), 859–63. DOI: 10.1007/s000180050048 46

Nerlich, A.G., Schleicher, E. D., Boos, N. (1997). Volvo award winner in basic science studies. Immunohistologic markers for age-related changes of human lumbar intervertebral discs. *Spine* 22(24), 2781–95. DOI: 10.1097/00007632-199712150-00001 48

Nerurkar, N. L., Baker, B. M., Sen, S., Wible, E. E., Elliott, D. M., Mauck, R. L. (2009). Nanofibrous biologic laminates replicate the form and function of the annulus fibrosus. *Nat Mater* 8(12), 986–92. DOI: 10.1038/nmat2558 59

Nesti, L. J., Li, W. J., Shanti, R. M., Jiang, Y. J., Jackson, W., Freedman, B. A., Kuklo, T. R., Giuliani, J. R., Tuan, R. S. (2008). Intervertebral disc tissue engineering using a novel hyaluronic acid-nanofibrous scaffold (HANFS) amalgam. *Tissue Eng Part A* 14(9), 1527–37. DOI: 10.1089/ten.tea.2008.0215 65

Nettles D. L., Vail T. P., Morgan M. T., Grinstaff M. W., Setton L. A. (2004). Photocrosslinkable hyaluronan as a scaffold for articular cartilage repair. *Ann Biomed Eng* 32, 391–7. DOI: 10.1023/B:ABME.0000017552.65260.94 65

Nettles, D. L., Chilkoti, A., Setton, L. A. (2009). Early metabolite levels predict long-term matrix accumulation for chondrocytes in elastin-like polypeptide biopolymer scaffolds. *Tissue Eng Part A* 15(8), 2113–21. DOI: 10.1089/ten.tea.2008.0448

Nettles, D. L., Haider, M. A., Chilkoti, A., Setton, L. A. (2010). Neural network analysis identifies scaffold properties necessary for *in vitro* chondrogenesis in elastin-like polypeptide biopolymer scaffolds. *Tissue Eng Part A* 2010, 16(1), 11–20. DOI: 10.1089/ten.tea.2009.0134

Nettles, D. L., Kitaoka, K., Hanson, N. A., Flahiff, C. M., Mata, B. A., Hsu, E. W., Chilkoti, A., Setton, L. A. (2008). In situ crosslinking elastin-like polypeptide gels for application to articular cartilage repair in a goat osteochondral defect model. *Tissue Eng Part A* 14(7), 1133–40. DOI: 10.1089/ten.tea.2007.0245

Neuenschwander, S., Hoerstrup, S. P. (2004). Heart valve tissue engineering, *Transpl Immunol* 12, 359–365. DOI: 10.1016/j.trim.2003.12.010

Nguyen K. T., West J. L. (2002). Photopolymerizable hydrogels for tissue engineering applications. *Biomaterials* 23, 4307–14. DOI: 10.1016/S0142-9612(02)00175-8 54

Nishida, K., Doita, M., Takada, T., Kakutani, K., Miyamoto, H., Shimomura, T., Maeno, K., Kurosaka, M. (2006). Sustained transgene expression in intervertebral disc cells *in vivo* mediated by microbubble enhanced ultrasound gene therapy. *Spine* 31(13), 1415–9. DOI: 10.1097/01.brs.0000219945.70675.dd

Nishida, K., Kang, J. D., Gilbertson, L. G., Moon, S. H., Suh, J. K., Vogt, M. T., Robbins, P. D., and Evans, C. H. (1999). Modulation of the biologic activity of the rabbit intervertebral disc by gene therapy: an *in vivo* study of adenovirus-mediated transfer of the human transforming growth factor-beta 1 encoding gene. *Spine* 24(23), 2419–25. 68

Norcross, J. P., Lester, G. E., Weinhold, P., Dahners, L. E. (2003). An *in vivo* model of degenerative disc disease. *J Orthop Res* 21(1), 183–8, DOI: 10.1016/S0736-0266(02)00098-0 49

Nordin, M., Weiner, S. S. (2001). *In*: Basic Biomechanics of the Musculoskeletal System, (Lindh M, ed.) Lippincott Williams & Wilkins Company New York, pp. 256–284. 46

O'Halloran, D., White, M., Collighan, R., Griffin, M., Pandit, A. (2005). Scaffold characterisation for nucleus pulposus regeneration. Abstract presented at the European Cells and Materials VI/SRN I Spinal Motion Segment: From Basic Science to Clinical Application, Davos, Switzerland, 68

O'Neill, C. W., Kurgansky, M. E., Derby, R., Ryan, D. P. (2002). Disc stimulation and pattern of referred pain. *Spine* 27(24), 2776–81. DOI: 10.1097/00007632-200212150-00007 43

O'Rorke, S., Keeney, M., Pandit, A. (2010). Non-viral polyplexes: scaffold mediated delivery for gene therapy. *Progress in Polymer Science* 35(4), 441–58. DOI: 10.1016/j.progpolymsci.2010.01.005

Oegema, T. R. Jr. (1993). Biochemistry of the intervertebral disc. *Clin. Sports Med.* 12, 419–439. 45

O'Halloran, D. M., Pandit, A. S. (2007). Tissue-engineering approach to regenerating the intervertebral disc. *Tissue Eng* 13(8), 1927–54. DOI: 10.1089/ten.2005.0608 68

Ohshima, H., Urban, J. (1992). The effect of lactate and pH on proteoglycan and protein synthesis rates in the intervertebral disc. *Spine* 17(9), 1079–82. DOI: 10.1097/00007632-199209000-00012 49

Ohshima, H., Urban, J. P., Bergel, D. H. (1995). Effect of static load on matrix synthesis rates in the intervertebral disc measured *in vitro* by a new perfusion technique. *J Orthop Res* 13(1), 22–29. DOI: 10.1002/jor.1100130106 46

Oki, S., Matsuda, Y., Shibata, T., Okumura, H., Desaki, J. (1996). Morphologic differences of the vascular buds in the vertebral endplate: scanning electron microscopic study. *Spine* 21(2), 174–7. DOI: 10.1097/00007632-199601150-00003 45

Osborne, J. L., Farmer, R., Woodhouse, K. A. (2008). Self-assembled elastin-like polypeptide particles. *Acta Biomater* 4, 49–57. DOI: 10.1016/j.actbio.2007.07.007

Pachence, J. M. (1996). Collagen-based devices for soft tissue repair. *J. Appl. Biomat.* 33, 35–40. DOI: 10.1002/(SICI)1097-4636(199621)33:1%3C35::AID-JBM6%3E3.0.CO;2-N

Pachence, J. M., Bohrer, M. P., Kohn, J. (2007). Biodegradable polymer. *In*: Principal of Tissue engineering, Lanza, R., Langer, R., Vacanti, J. eds.), 3rd edition, Academic Press, 323–39. 52

Palsson, B.O., Bhatia, S.N. (2004). Tissue Engineering (1st ed.), Pearson Prentice Hall, New Jersey, pp 35–55.

Peppas, N.A., Langer R. (1994). New challenges in biomaterials. *Science* 263(5154), 1715–20. DOI: 10.1126/science.8134835

Pfeiffer, M., Boudriot, U., Pfeiffer, D., Ishaque, N., Goetz, W., Wilke, A. (2003). Intradiscal application of hyaluronic acid in the non-human primate lumbar spine: radiological results. *Eur Spine J* 12(1), 76–83. 67

Rannou, F., Corvol, M., Revel, M., Poiraudeau, S. (2001). Disk degeneration and disk herniation: the contribution of mechanical stress. *Joint Bone Spine* 68(6), 543–6. DOI: 10.1016/S1297-319X(01)00325-6 49

Rannou, F., Revel, M., Poiraudeau, S. (2003). Is degenerative disk disease genetically determined? *Joint Bone Spine* 70(1), 3–5. DOI: 10.1016/S1297-319X(02)00003-9 49

Rastogi, A., Thakore, P., Leung, A., Benavides, M., Machado, M., Morschauser, M. A., Hsieh A. H. (2009). Environmental regulation of notochordal gene expression in nucleus pulposus cells. *J Cell Physiol* 220(3), 698–705. DOI: 10.1002/jcp.21816 60

Revell, P. A., Damien, E., Di Silvio, L., Gurav, N., Longinotti, C., Ambrosio, L. (2007). Tissue engineered intervertebral disc repair in the pig using injectable polymers. *J Mater Sci Mater Med.* 18(2), 303–8. DOI: 10.1007/s10856-006-0693-6 67

Reza, A. T., Nicoll, S. B. (2010)b). Characterization of novel photocrosslinked carboxymethyl-cellulose hydrogels for encapsulation of nucleus pulposus cells. *Acta Biomater* 6(1), 179–86. DOI: 10.1016/j.actbio.2009.06.004 66

Richardson, S. M., Hughes, N., Hunt, J. A., Freemont, A. J., Hoyland, J. A. (2008). Human mesenchymal stem cell differentiation to NP-like cells in chitosan-glycerophosphate hydrogels. *Biomaterials* 29(1), 85–93. DOI: 10.1016/j.biomaterials.2007.09.018

Roberts, S. (1991). Volvo award in basic sciences. Collagen types around the cells of the intervertebral disc and cartilage end plate: an immunolocalization study. *Spine* 16(9), 1030–8. DOI: 10.1097/00007632-199109000-00003 46

Roberts, S., Menage, J., Urban, J. (1989). Biochemical and structural properties of the cartilage end-plate and its relation to the intervertebral disc. *Spine* 14, 166–74. DOI: 10.1097/00007632-198902000-00005 44

Rong, Y., Sugumaran, G., Silbert, J., Spector, M. (2002). Proteoglycans synthesised by canine intervertebral disc cells grown in a type I collagen-glycosaminoglycan matrix. *Tissue Eng* 8, 1037–42. DOI: 10.1089/107632702320934137 65

Ruan, D., Xin, H., Zhang, C., Wang, C., Xu, C., Li, C., He, Q. (2010). Experimental intervertebral disc regeneration with tissue engineered composite in a canine model. *Tissue Eng Part A* (Epub ahead of print). DOI: 10.1089/ten.tea.2009.0770 67

Sagi, H., Bao, Q., Yuan, H. (2003). Nuclear replacement strategies. *Orthop Clin North Am* 34(2), 263–7. DOI: 10.1016/S0030-5898(03)00007-5 49

Sakai, D., Mochida, J., Iwashina, T., Hiyama, A., Omi, H., Imai, M., Nakai, T., Ando, K., Hotta, T. (2006). Regenerative effects of transplanting mesenchymal stem cells embedded in atelocollagen to the degenerated intervertebral disc. *Biomaterials* 27(3), 335–45. DOI: 10.1016/j.biomaterials.2005.06.038 59, 60, 67

Sakai, D., Mochida, J., Yamamoto, Y., Nomura, T., Okuma, M., Nishimura, K., Nakai, T., Ando, K., Hotta, T. (2003). Transplantation of mesenchymal stem cells embedded in Atelocollagen gel to the intervertebral disc: a potential therapeutic model for disc degeneration. *Biomaterials* 24(20), 3531–41. DOI: 10.1016/S0142-9612(03)00222-9 67

Sato, M., Asazuma, T., Ishihara, M., Kikuchi, T., Kikuchi, M., Fujikawa, K. (2003). An experimental study of the regeneration of the intervertebral disc with an allograft of cultured annulus fibrosus cells using a tissue-engineering method. *Spine* 28(6), 548–53. DOI: 10.1097/00007632-200303150-00007 65

Sawamura, K., Ikeda, T., Nagae, M., Okamoto, S., Mikami, Y., Hase, H., Ikoma, K., Yamada, T., Sakamoto, H., Matsuda, K., Tabata, Y., Kawata, M., Kubo, T. (2009). Characterization of *in vivo* effects of platelet-rich plasma and biodegradable gelatin hydrogel microspheres on degenerated intervertebral discs. *Tissue Engineering Part A* 15(12), 3719–27.

Schollmeier, G., Lahr-Eigen, R., Lewandrowski, K. (2000). Observations on fiber-forming collagens in the annulus fibrosus. *Spine* 25(21), 2736–41. DOI: 10.1097/00007632-200011010-00004 44

Schwarzer, A. C., Aprill, C. N., Derby, R., Fortin, J., Kine, G., Bogduk, N. (1995). The prevalence and clinical features of internal disc disruption in patients with chronic low back pain. *Spine* 20(17), 1878–83. DOI: 10.1097/00007632-199509000-00007 43

Sebastine I. M., Williams D. J. (2007). Current developments in tissue engineering of nucleus pulposus for the treatment of intervertebral disc degeneration. *Conf Proc IEEE Eng Med Biol Soc.* 2007, 6401–6. 44

Setton, L. A., Chen J. (2004). Cell mechanics and mechanobiology in the intervertebral disc. *Spine* 29, 2710–23. DOI: 10.1097/01.brs.0000146050.57722.2a 44

Sha'ban, M., Kim, S. H., Idrus, R. B. H., Khang G. (2008). Fibrin and poly(lactic-co-glycolic acid) hybrid scaffold promotes early chondrogenesis of articular chondrocytes: an *in vitro* study. *J Orthopaedic Surg Res* 3, 17. DOI: 10.1186/1749-799X-3-17 54, 59

Shalaby, S.W., DuBose, J.A., Shalaby, M. (2004). Chitosan based systems. In: Absorbable and Biodegradable Polymers (Shalaby, S.W.B. and Karen J.L., eds.), CRC Press, Boca Raton, FL, pp 79–89.

Shamji, M. F., Whitlatch, L., Friedman, A. H., Richardson, W. J., Chilkoti, A., Setton, L. A. (2008). An injectable and *in situ*-gelling biopolymer for sustained drug release following perineural administration. *Spine* 33(7), 748–54.

Shim, C. S., Lee, S. H., Park, C. W., Choi, W. C., Choi, G., Choi, W. G., Lim, S. R., Lee, H.Y. (2003). Partial disc replacement with the PDN prosthetic disc nucleus device: early clinical results. *J Spinal Disord Tech* 16(4), 324–30. 50

Smeds, K. A., Pfister-Serres, A., Miki, D., Dastgheib, K., Inoue, M., Hatchell, D. L., Grinstaff, M. W. (2001). Photocrosslinkable polysaccharides for *in situ* hydrogel formation. *J Biomed Mater Res* 54, 115–21. DOI: 10.1002/1097-4636(200101)54:1%3C115::AID-JBM14%3E3.0.CO;2-Q 65

Smelcerovic, A., Knezevic-Jugovic, Z., Petronijevic, Z. (2008) Microbial polysaccharides and their derivatives as current and prospective pharmaceuticals. *Curr Pharm Des* 14(29), 3168–95. DOI: 10.2174/138161208786404254

Snook, S. H. (2004). Work-related low back pain: secondary intervention. *J Electromyogr Kinesiol* 14, 153–60. DOI: 10.1016/j.jelekin.2003.09.006 43

Sobajima, S., Kim, J. S., Gilbertson, L. G., Kang, J. D. (2004). Gene therapy for degenerative disc disease. *Gene Ther* 11(4), 390–401. DOI: 10.1038/sj.gt.3302200

Stern, S., Lindenhayn, K., Schultz, O., Perka, C. (2000). Cultivation of porcine cells from the nucleus pulposus in a fibrin/hyaluronic acid matrix. *Acta Orthop Scand* 71(5), 496–502. DOI: 10.1080/000164700317381207 67

Su, W. Y, Chen, Y. C, Lin, F. H. (2010). Injectable oxidized hyaluronic acid/adipic acid, di-hydrazide hydrogel for nucleus pulposus regeneration. *Acta Biomater* (Epub ahead of print). DOI: 10.1016/j.actbio.2010.02.037 65

Sun, J. H., Zheng Q. X. (2009). Experimental study on self-assembly of KLD-12 peptide hydrogel and 3-D culture of MSC encapsulated within hydrogel *in vitro*. *Journal of Huazhong University of Science and Technology-Medical Sciences* 29(4), 512–16. DOI: 10.1007/s11596-009-0424-6

Sun, Y., Hurtig, M., Pilliar, R.M., Grynpas, M., Kandel, R.A. (2001). Characterization of nucleus pulposus-like tissue formed *in vitro*. *J Orthop Res* 19(6), 1078–84. DOI: 10.1016/S0736-0266(01)00056-0 44

Sztrolovics, R., Alini, M., Mort, J. S., Roughley, P. J. (1999). Age-related changes in fibromodulin and lumican in human intervertebral discs. *Spine* 24(17), 1765–71. DOI: 10.1097/00007632-199909010-00003 46

Takegami, K., An, H.S., Kumano, F., Chiba, K., Thonar, E.J., Singh, K., Masuda, K. (2005). Osteogenic protein-1 is most effective in stimulating nucleus pulposus and annulus fibrosus cells to repair their matrix after chondroitinase ABC induced *in vitro* chemonucleolysis. *Spine J* 5(3), 231–8. DOI: 10.1016/j.spinee.2004.11.001

Tanzer, M. L., Kimura, S. (1988). Phylogenetic aspects of collagen structure and function. *In:* Collagen (Nimni, M. E. ed.), CRC Press, Boca Raton, FL, pp. 55–98.

Thomas, J., Gomes, K., Lowman, A., Marcolongo, M. (2004). The effect of dehydration history on PVA/PVP hydrogels for nucleus pulposus replacement. *J Biomed Mater Res B Appl Biomater* 69(2), 135–40. DOI: 10.1002/jbm.b.20023

Thomas, J., Lowman, A., Marcolongo, M. (2003). Novel associated hydrogels for nucleus pulposus replacement. *J Biomed Mater Res A* 67(4), 1329–37 . DOI: 10.1002/jbm.a.10119 54

Thonar, E., An, H., Masuda, K. (2002). Compartmentalization of the matrix formed by nucleus pulposus and annulus fibrosus cells in alginate gel. *Biochemical Society Transactions* 30, 874–78.

Toolan, B. C., Frenkel, S. R., Yalowitz, B. S., Pachence, J. M., Alexander, H. (1996). An analysis of a collagen–chondrocyte composite for cartilage repair. *J Biomed Mat Res* 31, 273–80. 60

Trout, J. J., Buckwalter, J. A., Moore, K. C., Landas, S. K. (1982a). Ultrastructure of the human intervertebral disc. I : changes in notochordal cells with age. *Tissue Cell* 14(2), 359–69. DOI: 10.1016/0040-8166(82)90033-7 47

Trout, J., Buckwalter, J., Moore, K. (1982b). Ultrastructure of the human intervertebral disc II: cells of the nucleus pulposus. *Anat Rec* 204(4), 307–14. DOI: 10.1002/ar.1092040403 47

Turgut, M., Uysal, A., Uslu, S., Tavus, N., Yurtseven, M. (2003). The effects of calcium channel antagonist nimodipine on end-plate vascularity of the degenerated intervertebral disc in rats. *J Clin Neurosci* 10(2), 219–23. DOI: 10.1016/S0967-5868(02)00336-3 49

Urban JP, Holm S, Maroudas A, Nachemson A. (1982). Nutrition of the intervertebral disc: effect of fluid flow on solute transport. *Clin Orthop Relat Res* 170, 296–302. 46

Urban, J. P., Smith, S., Fairbank J. C. (2004). Nutrition of the intervertebral disc. *Spine* 29, 2700–09. DOI: 10.1097/01.brs.0000146499.97948.52 46

Urban, J., Maroudas, A. (1980). The chemistry of the intervertebral disc in relation to its physiological function. *Clin Rheumatol Dis* 6, 51–73. 46

Urban, J., McMullin, J. F. (1985). Swelling pressure of the intervertebral disc: influence of proteoglycan and collagen contents. *Biorheology* 22, 145–57. 46

Urban, J., Roberts, S., Ralphs, J. (2000). The nucleus of the intervertebral disc from development to degeneration. *Am Zool* 40, 53–63.
DOI: 10.1668/0003-1569(2000)040[0053:TNOTID]2.0.CO;2 46, 47

van der Rest, W. J., Dublet, B., Champliaud, M. (1990). Fibril associated collagens. *Biomaterials* 11, 28–31.

Videman, T., Leppavuori, J., Kaprio, J., Battie, M.C., Gibbons, L.E., Peltonen, L., Koskenvuo, M. (1998). Intragenic polymorphisms of the vitamin D receptor gene associated with intervertebral disc degeneration. *Spine* 23(23), 2477–85. DOI: 10.1097/00007632-199812010-00002 49

Viñas-Castells, R., Holladay, C., di Luca, A., Díaz, V. M., Pandit, A. (2009). Snail1 down-regulation using small interfering RNA complexes delivered through collagen scaffolds. *Bioconjug Chem* 20(12), 2262–9. DOI: 10.1021/bc900241w

Waddell, G. (1996). Low back pain: a twentieth century health care enigma. *Spine* 21(24), 2820–25. DOI: 10.1097/00007632-199612150-00002 43

Wadhwa, S., Paliwal, R., Paliwal, S. R., Vyas, S. P. (2009). Chitosan and its role in ocular therapeutics. *Mini Rev Med Chem* 9(14), 1639–47. DOI: 10.2174/138955709791012292

Wakatsuki, T., Elson, E. L. (2003). Reciprocal interactions between cells and extra-cellular matrix during remodeling of tissue constructs. *Biophys Chem* 100, 593–99. DOI: 10.1016/S0301-4622(02)00308-3 54

Wallach, C. J., Kim, J. S., Sobajima, S., Lattermann, C., Oxner, W. M., McFadden, K., Robbins, P. D., Gilbertson, L. G., and Kang, J. D. (2006). Safety assessment of intradiscal gene transfer: a pilot study. *Spine* 6(2), 107–12. DOI: 10.1016/j.spinee.2005.05.002

Walmsley, R. (1953). The development and growth of the intervertebral disc. *Edinburgh Med J* 60, 341–64. 44, 47

Wan, Y., Feng, G., Shen, F. H., Laurencin, C. T., Li, X. (2008). Biphasic scaffold for annulus fibrosus tissue regeneration. *Biomaterials* 29(6), 643–52. DOI: 10.1016/j.biomaterials.2007.10.031

Wang J. Y., Baer A. E., Kraus V. B., Setton L. A. (2001). Intervertebral disc cells exhibit differences in gene expression in alginate and monolayer culture. *Spine* 26, 1747–51. DOI: 10.1097/00007632-200108150-00003 59

Wang, B. H., Campbell, G. (2009). Formulations of polyvinyl alcohol cryogel that mimic the biomechanical properties of soft tissues in the natural lumbar intervertebral disc. *Spine* 34(25), 2745–53. DOI: 10.1097/BRS.0b013e3181b4abf5 54

Wang, E., Overgaard, S. E., Scharer, J. M., Bols, N. C., Moo-Young, M. (1988). Occlusion immobilization of hybridoma cells. *In:* Chitosan - Biotechnology Techniques, pp. 133–136. DOI: 10.1007/BF01876164

Wehling, P., Schulitz, K. P., Robbins, P. D., Evans, C. H., Reinecke, J. A. (1997). Transfer of genes to chondrocytic cells of the lumbar spine. Proposal for a treatment strategy of spinal disorders by local gene therapy. *Spine* 22(10), 1092–97. DOI: 10.1097/00007632-199705150-00008 68

Wilke, H. J., Kavanagh, S., Neller, S., Haid, C., Claes, L. E. (2001). Effect of a prosthetic disc nucleus on the mobility and disc height of the L4–5 intervertebral disc postnucleotomy. *J Neurosurg* 95(2 Suppl), 208-14. DOI: 10.3171/spi.2001.95.2.0208 50

Wolfe, H.J., Putschar, W.G., Vickery, A.L. (1965). Role of the notochord in human intervertebral disk. I. Fetus and infant. *Clin Orthop* 39, 205–12. 47

Xie, L. W., Fang, H., Chen, A. M., Li, F. (2009). Differentiation of rat adipose tissue-derived mesenchymal stem cells towards a nucleus pulposus-like phenotype in vitro. *Chin J Traumatol* 12(2), 98–103. DOI: 10.3760/cma.j.issn.1008-1275.2009.02.007 60

Xu, J. W., Johnson, T. S., Motarjem, P. M., Peretti, G. M., Randolph, M. A., Yaremchuk, M. J., (2005). Tissue-engineered flexible ear-shaped cartilage, *Plast Reconstr Surg* 115, 1633–41. DOI: 10.1097/01.PRS.0000161465.21513.5D

Yamamoto, M., Ikada, Y., Tabata, Y. (2001). Controlled release of growth factors based on biodegradation of gelatin Hydrogel. *J Biomater Sci Polym Ed* 12, 77–88. DOI: 10.1163/156856201744461 65

Yang, L., Kandel, R. A., Chang, G., Santerre, J. P. (2009). Polar surface chemistry of nanofibrous polyurethane scaffold affects annulus fibrosus cell attachment and early matrix accumulation. *J Biomed Mater Res A* 91(4), 1089–99. DOI: 10.1002/jbm.a.32331 59

Yang, S. H., Chen, P. Q., Chen, Y. F., Lin, F. H. (2005). Gelatin/chondroitin-6-sufate copolymer scaffold for culturing human nucleus pulposus cells *in vitro* with production of extracellular matrix. *J Biomed Mater Res Part B-App Biomat* 74(1), 488–94. DOI: 10.1002/jbm.b.30221 54

Yang, S. H., Wu, C. C., Shih, T. T., Chen, P. Q. Lin, F. H. (2008). Three-dimensional culture of human nucleus pulposus cells in fibrin clot: comparisons on cellular proliferation and matrix synthesis with cells in alginate. *Artif Organs* 32(1), 70–3. DOI: 10.1111/j.1525-1594.2007.00458.x 59

Yao, C. H., Liu, B. S., Hsu, S. H., Chen Y. S., Tsai, C. C. (2004). Biocompatibility and biodegradation of a bone composite containing tricalcium phosphate and genipin crosslinked gelatin, *J Biomed Mater Res* 69A, 709–17. DOI: 10.1002/jbm.a.30045 65

Yao, J., Turteltaub, S. R., Ducheyne, P. (2006). A three-dimensional nonlinear finite element analysis of the mechanical behavior of tissue engineered intervertebral discs under complex loads. *Biomaterials* 27(3), 377–87. DOI: 10.1016/j.biomaterials.2005.06.036 67

Yasuma, T., Koh, S., Okamura, T., Yamauchi, Y. (1990). Histological changes in aging lumbar intervertebral discs: their role in protrusions and prolapses. *J Bone Joint Surg Am* 72(2), 220–9. 43

Yoon, S. T. (2005). Molecular therapy of the intervertebral disc. *Spine J* 5(6 Suppl.), 280S–286S. DOI: 10.1016/j.spinee.2005.02.017 68

Yoon, S. T., Park, J. S., Kim, K. S., Li, J., Attallah-Wasif, E. S., Hutton, W. C., and Boden, S. D. (2004). ISSLS prize winner: LMP-1 upregulates intervertebral disc cell production of proteoglycans and BMPs *in vitro* and *in vivo*. *Spine* 29(23), 2603–11. DOI: 10.1097/01.brs.0000146103.94600.85 68

Yu, J. (2002). Elastic tissues of the intervertebral disc. *Biochem Soc Trans* 30(Pt 6), 848–52. DOI: 10.1042/BST0300848 47

Zeegers, W., Bohnen, L., Laaper, M., Verhaegen, M. (1999). Artificial disc replacement with the modular type SB Charite' III. 2-year results in 50 prospectively studied patients. *Eur Spine J* 8(3), 210–17. DOI: 10.1007/s005860050160 50

Zeiter, S., der Werf, M., Ito, K. (2009). The fate of bovine bone marrow stromal cells in hydrogels: a comparison to nucleus pulposus cells and articular chondrocytes. *J Tissue Eng Regen Med.* 3(4), 310–20. DOI: 10.1016/S0021-9290(08)70130-1 59, 60

Zhang, Y., Li, Z., Thonar, E.J., An, H. S., He, T. C., Pietryla, D., Phillips, F. M. (2005). Transduced bovine articular chondrocytes affect the metabolism of cocultured nucleus pulposus cells *in vitro*: Implications for chondrocyte transplantation into the intervertebral disc. *Spine* 30(23), 2601–7. DOI: 10.1097/01.brs.0000187880.39298.f0

Zigler, J. (2003). Clinical results with ProDisc. European experience and U.S. investigation device exemption study. *Spine* 28(20), S163–6. DOI: 10.1097/00007632-200310151-00009 50

Authors' Biographies

SUNIL MAHOR

Sunil Mahor is currently a Post-doctoral researcher at the Network of Excellence for Functional Biomaterials (NFB) laboratory at the National University of Ireland Galway. He completed his PhD in Pharmaceutics in 2006 at the Dr. H. S. Gour University Sagar (India). Sunil's current research interests include gene therapy based IVD regeneration strategies, vaccine development, advanced drug and gene delivery using synthetic nanocarriers, nanoparticles and the development of ECM based biomaterial based drug delivery systems.

ESTELLE COLLIN

Estelle Collin is a researcher at the Network of Excellence for Functional Biomaterials (NFB) laboratory at the National University of Ireland, Galway. She completed her Professional Master of Biology Cellular and Molecular engineering at the Université des Sciences et Technologies de Lille (France) in 2008. Estelle's current research efforts involve the development of cell and gene therapy of the intervertebral discs using an *in-situ* cross-linkable cell-seeded scaffold for the regeneration of degenerated discs. She is also investigating the addition of siRNA into the reservoir which it is hypothesized will slow down the degeneration process by repressing the up-regulated genes implicated in this process in the resident cells.

BIRAJA DASH

Biraja Dash is currently a researcher at the Network of Excellence for Functional Biomaterials (NFB) laboratory at the National University of Ireland, Galway. He completed his Masters in Life Sciences in 2005 from Sambalpur University (India). He has also worked as a Junior Research Fellow at the Indian Institute of Technology, Kharagpur (India). His research interests includes the fabrication of elastin peptide PEG based platform hollow spheres which can respond locally to the pro-thrombotic response of injured coronary vasculature. This platform technology can also be adapted to specific requirements for treating chronic (disease-generated) or acute (procedure-mediated) injury.

DAVID EGLIN

David Eglin earned his PhD from Nottingham Trent University (UK) under the supervision of Professor Carole C. Perry in 2002 after which he joined the group of Professor Jacques Livage at the

College de France (FR). He is currently principal investigator in the Musculoskeletal Regeneration Program at the AO Research Institute Davos (CH). His research interests include cell and drug delivery systems and biodegradable polymers for regenerative medicine.

ABHAY PANDIT

Abhay Pandit is the Director of the Network of Excellence for Functional Biomaterials (NFB); a Science Foundation Ireland funded Strategic Research Cluster at the National University of Ireland, Galway. The NFB research programme hosts several technology platforms associated with the development of biomaterials for clinical applications. Functionality to these forms is achieved through custom chemistries which facilitate the attachment of surface tethered moieties or encapsulated therapeutic factors including drugs, genes and other active agents. Although these platforms have been developed for musculoskeletal, cardiovascular, soft tissue repair or neural targets, other clinical targets including cancer and diagnostics are yet to be exploited.

Printed in the United States
by Baker & Taylor Publisher Services